WITHDRAWN

D1145717

3 1111 01507 3289

OXFORD CHEMISTRY PRIMERS

Physical Chemistry Editor	Founding/Organic Editor	Inorganic Chemistry Editor	Chemical Engineering Editor
RICHARD G. COMPTON	**STEPHEN G. DAVIES**	**JOHN EVANS**	**LYNN F. GLADDEN**
University of Oxford	University of Oxford	University of Southampton	University of Cambridge

Foundations of Inorganic Chemistry

Mark J. Winter

Department of Chemistry, University of Sheffield

John E. Andrew

(retired) Head of Chemistry, Hall Cross Comprehensive School, Doncaster

Series sponsor: AstraZeneca

AstraZeneca is one of the world's leading pharmaceutical companies with a strong research base. Its skill and innovative ideas in organic chemistry and bioscience create products designed to fight disease in seven key therapeutic areas: cancer, cardiovascular, central nervous system, gastrointestinal, infection, pain control, and respiratory.

AstraZeneca was formed through the merger of Astra AB of Sweden and Zeneca Group PLC of the UK. The company is headquartered in the UK with over 50,000 employees worldwide. R&D centres of excellence are in Sweden, the UK, and USA with R&D headquarters in Södertälje, Sweden.

AstraZeneca is committed to the support of education in chemistry and chemical engineering.

OXFORD
UNIVERSITY PRESS

This book has been printed digitally and produced in a standard specification
in order to ensure its continuing availability

OXFORD
UNIVERSITY PRESS

Great Clarendon Street, Oxford OX2 6DP

Oxford University Press is a department of the University of Oxford.
It furthers the University's objective of excellence in research, scholarship,
and education by publishing worldwide in
Oxford New York

Auckland   Cape Town Dar es Salaam   Hong Kong   Karachi
Kuala Lumpur Madrid Melbourne Mexico City   Nairobi
New Delhi Shanghai Taipei Toronto
With offices in
Argentina Austria Brazil Chile Czech Republic France Greece
Guatemala Hungary Italy Japan South Korea Poland Portugal
Singapore Switzerland Thailand Turkey Ukraine Vietnam

Oxford is a registered trade mark of Oxford University Press
in the UK and in certain other countries

Published in the United States
by Oxford University Press Inc., New York

© Mark J. Winter and John E. Andrew, 2000

The moral rights of the author have been asserted

Database right Oxford University Press (maker)

Reprinted 2009

All rights reserved. No part of this publication may be reproduced,
stored in a retrieval system, or transmitted, in any form or by any means,
without the prior permission in writing of Oxford University Press,
or as expressly permitted by law, or under terms agreed with the appropriate
reprographics rights organization. Enquiries concerning reproduction
outside the scope of the above should be sent to the Rights Department,
Oxford University Press, at the address above

You must not circulate this book in any other binding or cover
And you must impose this same condition on any acquirer

ISBN 978-0-19-879288-8

Printed and bound in Great Britain by CPI Antony Rowe, Chippenham and Eastbourne

Series Editor's Foreword

Most Oxford Chemistry Primers are designed to give a concise introduction to all chemistry students by providing the material that would usually form an 8–10 lecture course. As well as providing up-to-date information, this series expresses the explanations and rationales that form the framework of current understanding of inorganic chemistry. This Primer is aimed at providing the fundamentals of inorganic chemistry that are appropriate to students beginning a university chemistry course. It is seen as the link for this area between university and school or sixth form college. As such it should appeal to students taking A-levels and preparing for university, and also to those taking first level university courses.

Mark Winter has already written two Primers which have both been very successful. The quality of their presentation and the clarity of their approach is reflected in this book. John Andrew is a highly experienced sixth form teacher. Their combined experience has resulted in a very approachable textbook covering the Foundations of Inorganic Chemistry.

John Evans
Department of Chemistry, University of Southampton

Preface

This book introduces some concepts of inorganic chemistry clearly and concisely. The material could form the basis of an introductory course in inorganic chemistry and is designed to be useful as a link between school and university. When commencing a chemistry course, many students find the amount and variety of material to be mastered somewhat bewildering. It is hoped that the reader will appreciate that there is some order amongst the apparent chaos.

Coverage is aimed at the *student* rather than the *lecturer*. Our aim is to develop an understanding and appreciation of some chemical ideas, and that these ideas will encourage the student to adapt and extend current models to new situations. It is anticipated that this short text could find a place *alongside* textbooks containing more detailed coverage.

Many people made constructive criticism during the preparation of this book, particularly Nigel Mason, and we acknowledge them here. The remaining errors and misconceptions are, of course, ours.

Sheffield   Mark Winter
July 2000   John Andrew

Contents

1 Elements and periodicity

Matter is made up from very small particles called *atoms*. To date, 109 different atom types are known and named. Each type is referred to as an *element*. Claims for new elements with atomic numbers 110–112, 114, 116, and 118 are being analysed. Most matter in the universe is made up of just two elements, hydrogen (about 75% by weight) and helium (about 23% by weight). Heavier elements are synthesized in stars by nuclear reactions and those present in our environment represent the debris from stellar explosions from long ago.

1.1 Atomic structure

Sometimes, it is useful to regard atoms as small billiard ball-like entities, but usually a chemist does need a more advanced view of atom structure. Atoms possess structure. They are not little, hard, featureless balls, although they are spherical. An atom consists of a positively charged *nucleus* surrounded by negatively charged *electrons*. Most of the volume of the atom is associated with the electrons. Although the radius of the nucleus is only about 0.01% of that of the atom, the total mass of the electrons is much less than that of the nucleus. Atoms are very small, with radii of the order of 100 pm, meaning that several million atoms could line up in a row less than a millimetre long.

The nucleus contains one or more protons which from the chemist's point of view are positively charged indivisible particles. Apart from hydrogen, the nucleus also contains neutrons, also indivisible, which weigh about the same as protons, but are electrically neutral. The identity of the atom is defined by the number of protons within the nucleus. So, the nucleus within all hydrogen atoms possesses one and only one proton, the nucleus within all helium atoms contain two protons, and so on. The nucleus is surrounded by a number of electrons. Electrons are negatively charged but the magnitude of the charge is precisely the same as that upon a proton. The effect is that the atom is electrically neutral.

Atomic number: number of protons

While all atoms of one element possess the same number of protons, the nuclei of atoms of any one element may contain different numbers of neutrons. Atoms with the same number of protons but differing number of neutrons are called *isotopes*. So, all carbon atoms possess six protons and most contain six neutrons. The atomic mass of this form of carbon is 12 (6 + 6) and this isotope is denoted $^{12}_{6}C$. A small number of carbon atoms (1.1% in naturally occurring carbon) contain seven neutrons. These have an atomic mass of 13 and are denoted $^{13}_{6}C$. A very small proportion of carbon atoms, perhaps 1 in 10^{12} within living creatures, possess eight neutrons. These carbon atoms are $^{14}_{6}C$. The nuclei of carbon-14 atoms are not stable and decompose slowly, with a half-life of 5715 years, to the nitrogen isotope $^{14}_{7}N$

Carbon-12: 6 protons + 6 neutrons
Carbon-13: 6 protons + 7 neutrons
Carbon-14: 6 protons + 8 neutrons

$$^{14}_{6}C \rightarrow ^{14}_{7}N + e^-$$

in a process which releases a β-particle (an electron) from the nucleus and γ-radiation. Isotopes which decompose are 'radioactive'. In this particular case, the radioactive decay is useful in that it forms the basis of radiocarbon dating.

The masses of the carbon isotopes weighted for their relative proportions in nature constitute the relative atomic mass. By international convention, relative atomic masses are expressed on a scale relative to the mass of the $^{12}_{6}C$ isotope (12.0000). Since the proportion of $^{13}_{6}C$ is only 1.1%, the relative atomic mass of carbon is slightly over 12.0000, actually 12.0107 according to the latest (1997) values. It is common to see relative atomic masses expressed as the term *atomic weight*, indeed the definitive IUPAC (International Union of Pure and Applied Chemistry) tables do exactly that.

Conceptually, one of the simplest models of atomic structure is the Bohr model. In the Bohr model of the hydrogen atom (Fig. 1.1), the single electron rotates around the nucleus in a 'planetary' fashion within an *orbit*. The radius of the orbit is denoted r. Under normal conditions, in *every* hydrogen atom, the electron revolves around the nucleus at the *same* distance r. The electron in motion possesses a certain amount of energy (E), which in practice can be calculated. This energy, E, depends upon the radius, r, of the orbit (that is, the energy, E, is *a function of* the radius, r). Since the radius is the same for every hydrogen atom, the energy of the electron is also the same for every hydrogen atom. The electron is said to reside at a particular energy level. Under certain circumstances it is possible, by inducing the atom to absorb some energy, for the electron to transfer to another orbit and so to a different energy level further from the nucleus (since the new radius is different from the first radius). A key feature of the Bohr description is that *only certain values of orbit radius are allowed*. From this it follows that only certain energy levels are allowed. The radius (and therefore the energy) is said to be *quantized*. Allowed radii (called *shells*) are labelled with quantum numbers 1, 2, 3... but the electron always resides in the level labelled 1 unless that atom is in an excited state (after promotion of the electron in some way). These particular quantum numbers are *principal quantum numbers*. They do not denote the radius, they are simply labels.

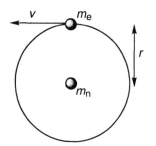

Fig. 1.1 The Bohr hydrogen atom.

The Bohr model of the hydrogen atom may be extended to multielectron elements starting from helium. The Bohr model of the fluorine atom is shown in Fig. 1.2. In all cases, this simple pictorial model has electrons revolving around the central nucleus in orbits at fixed distances. These pictures are useful at an elementary level in that they help to keep track of electrons; however; their usefulness is diminished because of the danger that they foster the notion that this is what atoms 'look like'.

The Bohr model is useful for many reasons, not the least of which is that it is easy to understand. However, it is fundamentally flawed. In the Bohr model, there seems nothing to stop the electrons spiralling into the nucleus. A useful discussion of many chemical phenomena requires a more sophisticated model. Unfortunately any more sophisticated model is inherently more difficult to understand, in part because of difficulties in visualizing it.

It is often useful to get away from the planetary electron model and to regard the electron as an altogether more tenuous entity smeared over a

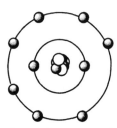

Fig. 1.2 Bohr model of the fluorine atom.

volume of space (Fig. 1.3). In this 'non-classical' view, it is not appropriate to refer to the electron as a charged particle that revolves around the nucleus. Instead, the electron in that region of space is represented by an *orbital*, a term clearly related to the word orbit. In Fig. 1.3, the *density* of the dots represents *electron density* in hydrogen's lowest energy orbital, otherwise known as the hydrogen 1*s* orbital. The diagram is actually a slice through the spherically symmetrical three-dimensional shape representing the orbital. The density falls away exponentially with distance from the nucleus. The electron density is related to the probability of finding the electron at any one point.

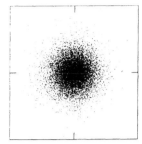

Fig. 1.3 The hydrogen 1*s* atomic orbital.

Fig. 1.4 Shorthand representations of the hydrogen 1*s* atomic orbital.

It is not always convenient to draw complicated density diagrams, which require computer programs to generate, so shorthand representations such as those in Fig. 1.4 are commonly employed.

As for the Bohr model, it is possible to calculate the energy associated with the hydrogen 1*s* orbital, and the energy calculated is *precisely* the same energy as associated with the Bohr orbit. Further, under the same conditions referred to above for the Bohr model, if the atom is induced to absorb some energy, then the electron is *promoted* from one orbital to another. Since the energy of the electron in the new orbital is different from that in the first, the electron has changed energy levels.

As in the Bohr model, only certain values of energy level are allowed. The energy is quantized. One important distinction between this model and the Bohr model, however, is that the quantization of the energy levels arises perfectly naturally from the mathematics used to describe the orbitals. In the Bohr model the quantization is an artificial constraint imposed to fit the observed data.

The orbital depicted in Fig. 1.3 is spherically symmetrical and is called an *s* orbital (for historical reasons). Its full name is 1*s*. The number 1 is a quantum number, in this case known as the *principal quantum number* (the same quantum number as in the Bohr model). The electronic structure of hydrogen is written as $1s^1$ ('one-*s*-one'). This tells us that there is a single electron (denoted by the superscript) in the *s*-orbital of the first shell.

Shell: a set of energy levels all of which have the same principal quantum number, *n*.

Any orbital can accommodate up to just *two* electrons, never more. Certain properties of electrons are treated conveniently as if the electrons spin about an axis passing through a diameter of the electron (the *particle model*), rather like the earth spins about an axis every 24 hours. When there are two electrons in an orbital, one is said to *rotate* or *spin* in one direction and the second in the other. This is unfortunate terminology perhaps, but in common usage. The *spin quantum number* differentiates between the two directions of spin and takes the values $+\frac{1}{2}$ or $-\frac{1}{2}$. This spin is not to be confused with the notion of electrons rotating about, or orbiting, the nucleus.

The hydrogen atom contains a single proton. The next element after hydrogen is helium with two protons — and so two electrons. The second electron also resides in the 1*s* orbital and the electronic structure of helium is written as $1s^2$ ('one-s-two').

Not all orbital functions are spherical. In fact, only *s* orbitals are spherically symmetrical. A second group of orbitals are called *p* orbitals. There are three distinct *p* orbitals. There are no *p* orbitals in the *n*=1 shell, but

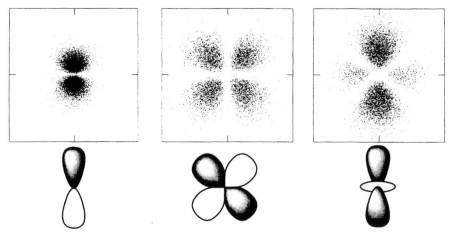

Fig. 1.5 Slices (top) though a $2p$ orbital (left) and two of the $3d$ orbital functions (centre and right). Schematic representations are shown beneath.

there are three in all other shells. Each of the three $2p$ orbitals contains two lobes (Fig. 1.5) and are respectively orientated along the x, y, and z axes. Again, the origin of the letter p is historical. The electron density is shown for one of the $2p$ orbitals in Fig. 1.5 (left). Orbitals known as d orbitals are more complex again. Four of the d orbital functions possess four lobes (Fig. 1.5, centre) and the fifth (the d_{z^2} orbital) has two lobes with a further ring of electron density surrounding those two lobes (Fig. 1.5, right).

The electronic structures of the first 18 elements are given in Table 1.1. The shorthand form in the final column becomes much more useful for heavy elements than for light elements.

Table 1.1 Electronic structures of elements 1–18 (gas phase)

Z	Symbol	Electronic configuration	Shorthand form
1	H	$1s^1$	$1s^1$
2	He	$1s^2$	$1s^2$
3	Li	$1s^22s^1$	$[\text{He}]2s^1$
4	Be	$1s^22s^2$	$[\text{He}]2s^2$
5	B	$1s^22s^22p^1$	$[\text{He}]2s^22p^1$
6	C	$1s^22s^22p^2$	$[\text{He}]2s^22p^2$
7	N	$1s^22s^22p^3$	$[\text{He}]2s^22p^3$
8	O	$1s^22s^22p^4$	$[\text{He}]2s^22p^4$
9	F	$1s^22s^22p^5$	$[\text{He}]2s^22p^5$
10	Ne	$1s^22s^22p^6$	$[\text{He}]2s^22p^6$
11	Na	$1s^22s^22p^63s^1$	$[\text{Ne}]3s^1$
12	Mg	$1s^22s^22p^63s^2$	$[\text{Ne}]3s^2$
13	Al	$1s^22s^22p^63s^23d^1$	$[\text{Ne}]3s^23p^1$
14	Si	$1s^22s^22p^63s^23p^2$	$[\text{Ne}]3s^23p^2$
15	P	$1s^22s^22p^63s^23p^3$	$[\text{Ne}]3s^23p^3$
16	S	$1s^22s^22p^63s^23p^4$	$[\text{Ne}]3s^23p^4$
17	Cl	$1s^22s^22p^63s^23p^5$	$[\text{Ne}]3s^23p^5$
18	Ar	$1s^22s^22p^63s^23p^6$	$[\text{Ne}]3s^23p^6$

1.2   The periodic table

The name most commonly associated with the development of the periodic table is Dmitri Mendeleev, a Russian chemist who published an important document in 1869. He described a periodic table and left gaps in it where he predicted that then unknown elements should occur. His predictions were based upon the observations of *periodicity* in the properties of the elements. However, other chemists made enormously important contributions as well. In particular, John Newlands, an English chemist, observed in 1864 that patterns existed in the properties of the elements. When the elements are arranged in order of atomic weight, he commented that the properties of the eighth element resembled those of the first, those of the ninth resemble those of the second, and so on. He divided the elements into periods and groups according to his 'law of octaves'. In 1864 a German chemist called Lothar Meyer published a partial periodic table and revised his table later in 1869. Much earlier (1817), a German chemist Johann Dobereiner noted that there were a number of groups of elements (which he called triads) such as calcium, strontium, and barium with related properties for which the atomic weight of the centre element is about the average of the first and last.

Originally, the periodic table was constructed as a way to organize the chemical elements (many of those known now were unknown at the time) into groups of elements with related chemical properties. For instance, elements forming oxides with the stoichiometric formula M_2O are grouped together. Elements forming oxides with the stoichiometric formula MO are placed into a different group. From a more modern perspective, one can appreciate that since chemical properties rely upon *electronic structure*, then a periodic table based upon elemental electronic structures is most appropriate. Perhaps the magnitude of the achievements of Newlands, Meyer, Mendeleev, and others in constructing the periodic table is appreciated more when it is realized that their work was carried out nearly half a century before the first even slightly sensible ideas concerning atomic structure were proposed by Rutherford.

Stoichiometry: the relative proportions of the elements in a given compound.

Over the years, many forms of the periodic table have been devised but the current standard table is that shown on the inside cover of this book. The standard form is not necessarily the best, but as it is the most commonly used it is important to be familiar with it. The elements in the standard table are organized into four blocks, *s*, *p*, *d*, and *f*. The *f*-block elements are placed as a distinct block below the other three blocks, not for any chemical reason but for ease of organization on the printed page. On moving from left to right across the *p* block, successive elements gain an additional electron until the *p* level is filled with the final element of that *p* block element row. The situation is related for the *s*, *d*, and *f* block elements. Helium ($1s^2$) does not have any *p* electrons but nevertheless is placed over neon rather than beryllium ($2s^2$) because in helium and neon, the final electron completes a *valence shell*. Hydrogen is placed over lithium as both have ns^1 configurations. However, hydrogen has the ability to accept a single electron to complete the valence shell, as does fluorine. Some tables therefore place

Valence shell: the outermost shell of electrons.

Table 1.2 Typical metal and non-metal physical properties

Metals	Non-metals
Shiny	Non-reflective
Hard	Soft
Good heat conductors	Poor heat conductors
Good electricity conductors	Poor electricity conductors
Malleable, ductile	Brittle
High tensile strength	Low tensile strength

Table 1.3 Distinctions between metal, non-metal (bold text), and metalloid (shaded boxes)

B	C	N	O	F	Ne
Al	Si	P	S	Cl	Ar
Ga	Ge	As	Se	Br	Kr
In	Sn	Sb	Te	I	Xe
Tl	Pb	Bi	Po	At	Rn

hydrogen in two groups, 1 and 17, to reflect this, or even as a lone element apart from the main body of the table. Note that a successful periodic table allows the user to determine the electronic structure of most if not all of the elements.

1.3 Periodicity

Periodic trends which apply to specific areas of the periodic table are discussed in later chapters. This section addresses a few general properties across the whole table.

One of the clearest classifications for the elements is into metals or non-metals. There are distinctions of metal and non-metal properties (Table 1.2) in addition to clear differences in chemistries. The reasons for this are associated with the abilities of the atoms of the elements to bond together and these are associated with their electronic structures. Metallic properties are more apparent to the left and to the bottom of the periodic table. In practice, a number of elements display properties somewhat intermediate between those of metal and non-metal. These are referred to as semi-metallic or metalloid. What constitutes semi-metallic is slightly open to interpretation but a diagonal set of elements (Table 1.3) across the *p*-block elements is referred to as metalloid. Some texts treat a slightly different set of elements as metalloid. Some elements show exceptional properties. So, while carbon as graphite is regarded as a non-metal, it conducts electricity well.

Clear trends in physical properties of the elements are apparent and illustrated for just one property, melting point, in Fig. 1.6. Metals tend to have high melting points although there is a large range. Some non-metals and metalloids (carbon, boron) have high melting points. Melting points tend to decrease down the group for Groups 1 and 2 and the *p*-block elements (with exceptions) but increase down the group for the *d*-block elements.

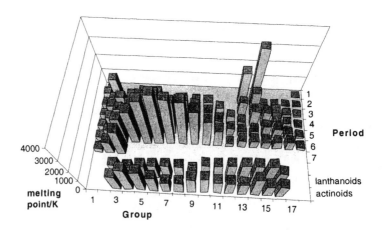

Fig. 1.6 Melting points of the elements across the periodic table.

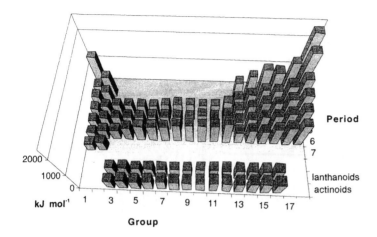

Fig. 1.7 Periodic table trends for the first ionization energy.

The abrupt transition between carbon and nitrogen is interesting. Nitrogen is able to satisfy the octet rule (Chapter 2) through the formation of a triple bond between two atoms. The resulting species, N_2, interacts relatively little with adjacent molecules and so the melting point is low. Carbon cannot satisfy the octet rule by forming a diatomic species since to do so would require a quadruple bond, not possible for *p*-block elements. It is able, however, to achieve an octet structure by making bonds in lattices, either as graphite or diamond (Section 5.7).

Some elements lose electrons more easily than others. Others show tendencies to acquire electrons. The ease by which electrons are lost from a neutral atom is measured by the first ionization energy. This is the energy change, in Eqn 1.1, and values of the first ionization energy are displayed in Figs. 1.7 and 1.8. It is also possible to measure second and subsequent ionization energies which are the corresponding energy changes in the formation of $M^{2+}(g)$ and so on.

$$M(g) \rightarrow M^+(g) + e^-(g)$$

The ease with which the outermost electron is removed depends upon the distance of the electron from the nucleus. This seems reasonable given the existence of an inverse square law governing the attraction of the nucleus to the surrounding electrons. There are other factors, however, including the charge on the nucleus and the effect of any other electrons in the atom or ion. The greater the nuclear charge, the more strongly held is the electron. However, the nuclear charge effectively is lessened, or *screened*, by any intervening electrons. Although the nuclear charge increases on descending a group in the periodic table, this is more than compensated for by the greater distance of the outermost electron from the nucleus and the screening effect

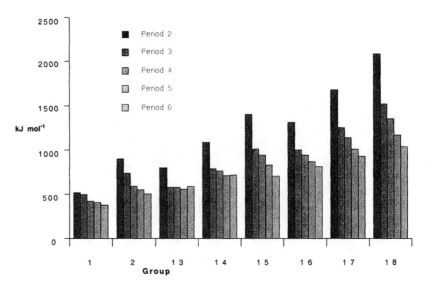

Fig. 1.8 First ionization energies of the *s*- and *p*-block elements plotted against group number.

of the inner electrons. The consequence is that ionization enthalpies fall on descending a group in the periodic table.

Ionization energies increase on moving from left to right along a row in the periodic table. On moving along the row, the nuclear charge increases, causing the atom's electrons to move close to the nucleus and the energy levels, including the outermost one, to drop in energy. The practical consequence of this is that electrons are removed far less easily to the right of the periodic table. Elements with low ionization energies lie to the left of the table and to the bottom of the table. The chemistry of these elements is dominated by tendencies to lose electrons.

The tendency for atoms of neutral elements to gain electrons is measured by the electron affinity. This is numerically equal to the negative of the electron gain enthalpy, which is the enthalpy change when a gas-phase atom gains an electron (shown in Eqn 1.2 for chlorine).

$$Cl(g) + e^-(g) \rightarrow Cl^-(g) \qquad E_{EA} = +349 \text{ kJ mol}^{-1}, \Delta H = -349 \text{ kJ mol}^{-1} \qquad (1.2)$$

Inspection of Fig. 1.9 shows that elements to the right and top of the periodic table show the greatest tendencies to gain electrons. This feature is extremely important in their chemistry.

Once in a molecule, there is no particular reason for the distribution of electrons in bonds between different atoms to be equal. Some atoms *in molecules* tend to attract bonded electrons far more strongly than others. This tendency is put on a quantitative basis by the *electronegativity* concept (see Fig. 1.10). Strongly electronegative elements in molecules strongly attract electrons. Weakly electronegative elements in molecules show little tendency to attract electrons. Such elements are sometimes referred to as electropositive.

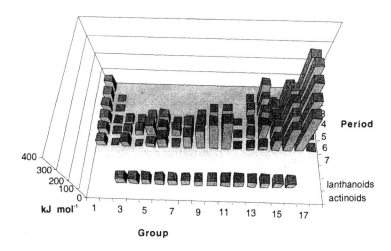

Fig. 1.9 Periodic table trends for the electron affinity.

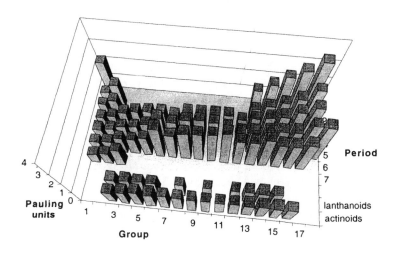

Fig. 1.10 Periodic table trends for Pauling electronegativity.

There are several electronegativity scales in use and the basis of each is different. Perhaps the best known is the Pauling scale (Fig. 1.10). This is based upon differences between observed and calculated vales of bond dissociation energies in molecules AA, BB, and AB. The figure demonstrates that the *most* electronegative elements are positioned to the *right* and *top* of the periodic table while the *least* electronegative are to the *bottom* and *left* of the table. In a molecule containing a bond A—B, if B is more electronegative than A, then the charge distribution in that bond is such that B is more negatively charged. This is important, since partial positive and negative

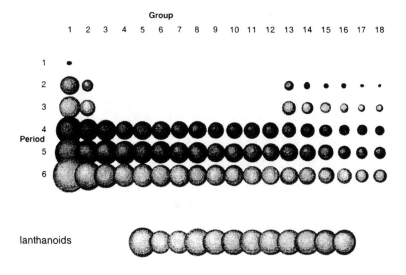

Fig 1.11 Atomic radii of the elements (calculated).

charges in molecules are important in defining at which points given reagents will attack.

Size is a fundamental property of atoms. It is not easy to assign values to size, however. Since electrons are contained within orbitals that do not have a definite limit, there is no precise edge to an atom isolated in space. It is therefore necessary to *define* size in some way. The size could be defined in molecules such as H_2 as one-half of the internuclear distance, the *single non-polar covalent radius*. Not all the elements form species M_2 containing a single M—M bond, O_2 and N_2 are examples. In such cases, a radius might be inferred from the lengths of bonds to elements of known size. There are other definitions of size. In metals, the *metallic radius* is one-half of the internuclear distance between two metal atoms. This measure of radius gives values that are not precisely the same as covalent radius. When the valence shells of atoms are in non-bonded contact, the non-bonded radius is one half of the internuclear distance. This radius is the van der Waals radius. It has great importance in the packing of atoms and molecules in crystals.

The size of an isolated atom can be *calculated* once an appropriate definition of size is defined. One such definition might be the distance between the nucleus and the distance of maximum electron density for the valence orbitals. These calculations are not trivial. The results of one such set of calculations are portrayed in Fig. 1.11. Size increases *down* the groups. Size decreases from left to right across the table. This is because increasing nuclear charge draws in the valence shell more tightly. Atoms to the left of table are larger because the inner electrons screen the valence electrons from the full nuclear charge. On moving across the periodic table from left to right, the size decreases as the increasing nuclear charge exerts an increasing effect through the core of inner electrons.

2 Bonding

Understanding what holds atoms together is not easy. What is clear, however, is that all bonding is a consequence of electrostatic interactions between positively charged nuclei and negatively charged electrons. The type of bonding that holds atoms together in a metal, such as iron, is called *metallic bonding*. Compounds such as salt (sodium chloride) consist of lattices of anions and cations held together via electrostatic forces. This is *ionic bonding*. Compounds such as oxygen and nitrogen in the air that we breathe are held together by shared-electron bonds. This is *covalent bonding*. Most compounds are best described as being somewhere between pure covalent and pure ionic in nature.

There are simple and less simple ways to describe bonding. Any attempt to describe bonding is called a model and even the most complicated models have problems. The Lewis model of bonding ('dot-and-cross') is credited to G.N. Lewis who worked in the area at the start of the twentieth century. His contribution to chemistry is immense, especially when it is recognized that virtually nothing was known about atomic structure at the time. Lewis noticed that some atoms towards the right of the periodic table form ions by gaining electrons in such a way that the total number of valence electrons surrounding that atom is eight (the 'octet' configuration). This is done either by electron donation to that atom (ionic bonding), or by electron sharing (covalent bonding) with one or more adjacent atoms. With the benefit of hindsight, it is clear that eight is the 'magic number' because it corresponds to the number of electrons required for filling all available energy levels in the valence shell, at least for the lighter elements up to neon.

The Lewis model has problems, as we shall see, but has the important advantage of simplicity. When the Lewis model of bonding fails in some way, perhaps the *hybridization model* will suffice. If and when that fails, maybe a further model involving *molecular orbitals* will be appropriate.

Some atoms are not held together by any bonds. Examples include the Group 18 noble gases. All the Group 18 gases freeze to form regular lattices of atoms on cooling. Since the solids exist, there must be some interaction between the atoms in these solids. These interactions are provided by weak attractive forces known as van der Waals forces.

2.1 Metallic structures

Whatever the nature of metal–metal bonding, there are interactions between adjacent metal atoms. Metals are interesting materials. They have high melting points, so clearly strong bonding must exist, and yet their shape is changed easily, suggesting that it is possible to move the atoms in the bulk metal relative to each other quite easily. There is more than one way to treat

The three bond types:
- metallic bonding
- ionic bonding
- covalent bonding

van der Waals forces: transient fluctuations in electron density on a molecule result in a dipole. This induces a fluctuation in electron density (and so a dipole) on a neighbouring molecule. The result is an attraction between the two instantaneously produced dipoles.

As metallic bonding involves loss of electrons to the 'sea' it is adopted by elements that have a low ionization energy. Suitable elements include the *s*- *d*- and *f*-blocks and a few of the *p*-block elements (see Chapter 5).

bonding in metals. One of the simplest is to envisage the metal as a *regular lattice array* of cations, each of which consists of the metal atom less the valence electrons. Those valence electrons are detached from any single cation and the spaces between the cations are filled with a 'sea' of electrons. The sea of electrons is mobile, accounting for the high electrical conductivity of metals. At a more sophisticated level, the behaviour of electrons in the 'sea' may be described using the *band level* model, but this is beyond the scope of this text.

When considering the solid state structures of metals it is convenient to regard these metal cores as small hard spheres. In order to describe the nature of metal packing, it is necessary to visualize the arrangement of the atoms or spheres in space. Atoms in solid metals are not arranged in random arrays. Instead, the atoms are arranged in regular repeating crystalline arrays. There are many different ways for the metal atoms cores to be arranged in regular lattice arrays.

One of the simplest crystalline arrays conceptually is a cubic array (Fig. 2.1). Perhaps surprisingly, this form is displayed only by a single metal — polonium.

In most, but not all cases, the total amount of bonding for a metal atom is maximized by placing as many metal atoms around it as possible. The problem therefore requires an analysis of the ways in which balls pack so as to maximize the total number of balls surrounding any one ball.

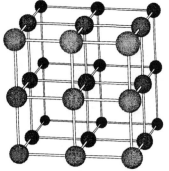

Fig. 2.1 The arrangement of atoms in elemental polonium.

One way to approach this problem is to arrange a number of balls in a tray so that as little space is taken up as possible. The result is the arrangement in the left section of Fig. 2.2. Note the hexagonal arrangement. Note also that each internal ball is surrounded by six others. The spheres are shown as not quite touching so that the diagrams might be a little clearer. The next step is to arrange a second layer of balls upon the first. In practice, the next layer of balls will fall naturally into place and this creates a second hexagonal array identical in appearance to the first but displaced (Fig. 2.2, centre). In order to distinguish these two layers, one is labelled *a* and the second is labelled *b*.

The third layer of balls can be arranged so that each ball is arranged directly over the balls in the first layer. The third layer is therefore labelled *a* as well and overall the three-layer structure is labelled *aba*. If the pattern is continued with more layers, the resulting structure is denoted *abababababababab*.... As the balls in this structure are arranged as tightly as possible, the packing is referred to as close packed. Because of the hexagonal

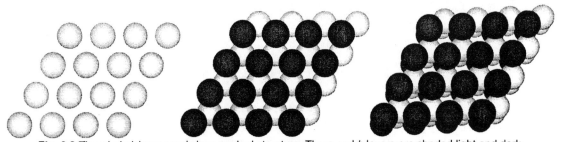

Fig. 2.2 The *ababab* hexagonal close-packed structure. The *a* and *b* layers are shaded light and dark.

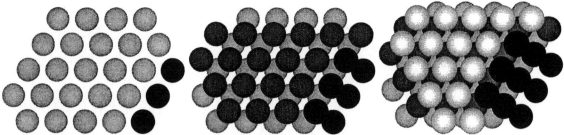

Fig. 2.3 The *abcabcabc* cubic close-packed structure.

nature of the packing, the structure is called *hexagonal close packed*. Note that each internal ball touches 12 other balls: the *coordination number* is 12.

There is another possibility however. The third layer can stack, as in Fig. 2.2 over the *a* layer to form the *abababab* hexagonal close-packed structure. Alternatively, the third layer can occupy a *third* position, *c*, by stacking over the holes evident in the centre structure of Fig. 2.2. The result is shown in the third part of Fig. 2.3. Following this, repeat layers are put down so that the resulting structure is labelled *abcabcabcabc*.... This type of structure is referred to as *cubic close-packing*. The *abcabc* structure from Fig. 2.3 is redisplayed in Fig. 2.4 after rotation and the spheres shaded dark show clearly the origin of the term cubic close packing.

Armed with a knowledge of internuclear distances in any given close-packed metal it is possible to define the metallic radius for that metal as one-half of that internuclear distance. Since the coordination number is 12 in close packed structures, this is sometimes referred to as the r_{12} for that metal.

The *abcabc* structure from Fig. 2.3 is rotated to show the origin of the term cubic close packing.

The *abababab* and *abcabc* structures are the most efficient in terms of the percentage space occupied per unit volume, but not all metals display one of these two structures in their normal form. A slightly less efficient form of packing known as body-centred cubic (Fig. 2.5) is displayed by the Group 1, 5, and 6 elements together with one or two other metals.

In this form of packing, instead of the first layer consisting of atoms surrounded by six other atoms in a hexagonal array, it consists of atoms surrounded by *four* other atoms in a square array. The second layer is displaced so as to fill in the pits and is also necessarily a square array. The third layer quite naturally fits in above the first layer making the layer structure *abababab* overall. Figure 2.5 shows a section of 16 atoms from this

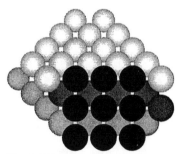

Fig. 2.4 The *abc* structure from Fig. 2.3 rotated to show the origin of the cubic terminology.

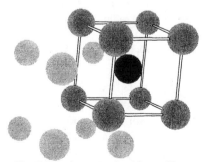

Fig. 2.5 Body-centred cubic packing.

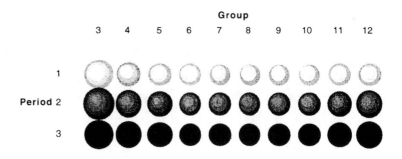

Fig. 2.6 Metallic radius (12 coordination) for the *d*-block elements.

structure and illustrates the origin of the term body centred cubic. The eight dark grey spheres are from adjacent layers of the *a* layer and form a perfect cube. The other eight spheres are from the *b* layer and the very dark sphere of that second set of atoms is located at the precise centre of the cube defined by the eight *a* atoms. Note that in this structure (edges excepted), every atom in the structure is at the centre of a cubic array of eight atoms.

Since the coordination number is 8 for such metals, it is not quite correct to compare directly a radius defined as one-half of the internuclear distance in body-centred cubic structures with the r_{12} values for close-packed metals as like is not being compared with like. Instead, values of radii obtained for body-centred cubic structures are converted so as to reflect the values they would have if 12-coordination was that displayed (Fig. 2.6). Note that the metallic radii tend to decrease towards the centre of the *d*-block. The first *d*-block period elements are smaller than the second and third series, which are about the same size as each other.

2.2 Ionic structures

Recall that Lewis noticed that some atoms towards the right of the periodic table form anions by gaining electrons in such a way that the total number of valence electrons is that of the next noble gas. Atoms of elements at the left of the periodic table form cations by losing electrons in such a way that the valence electron count is that of the previous noble gas. So in common salt, NaCl, chlorine gains an electron from sodium to form chloride and so achieve an octet, while sodium loses its valence electron to become Na⁺ (also an octet structure).

In one sense, ionic structures are related to metallic structures. Ionic compounds, such as salt, consist of arrays of atoms, in this case as cations and anions, in regular crystalline lattices.

Anions are generally larger than cations. Therefore, one can visualize the crystalline lattice as packed spherical anions. The cations are contained within the remaining spaces, or holes. Consider the structure of common salt, commonly displayed as in Fig. 2.7. Rather than drawing the diagram with the anions just touching each other, the spheres are scaled so as to aid

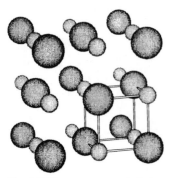

Fig. 2.7 The arrangement of atoms in the NaCl lattice.

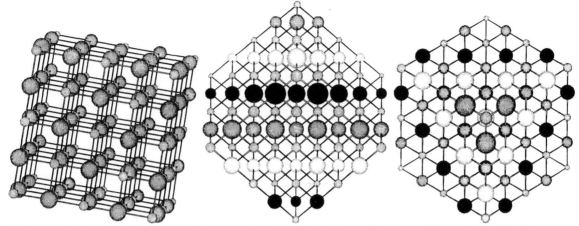

Fig. 2.8 The structure of salt, NaCl. The image on the left is the familiar presentation. while that on the right is the same image rotated and shaded to show the *abcabc...* cubic close-packed layers of chloride.

understanding. This lattice is shown in slightly extended form in Fig 2.8. Each of the components of Fig. 2.8 are the same piece of lattice, but displayed from different angles. The chloride ions are shaded in the second and third images to show that the lattice is made up as an *abcabcabc* close packed array of ions — in other words *cubic close-packed*. The sodium ions are smaller (116 pm) than the chloride ions (167 pm) and occupy the available space left after close packing of the chloride ions.

Determination of the size of an ion is not simple. While for a species consisting of a single element, such as a metal, it seems entirely reasonable to define the radius as one-half of the interatomic distance, this option is not available for any species containing more that one element. In any given interaction, where does one element start and the other finish? However, if the size of *one* of the ions is known then it is easy to calculate the size of the second. The way forward is to *define* the size of one anion (six-coordinate O^{2-} as 126 pm) and to use this value to determine the sizes of all the other ions based upon that single value.

The structure of the Group 1 halide NaCl is discussed above. However, not all halides show the same structure. If the size of the cation is much larger than that of the sodium cation, then there is insufficient room for the cation to fit within the holes in the cubic close-packed array of anions. This is the case for caesium (181 pm). Caesium chloride therefore adopts a different structure so as to accommodate the larger cation.

A measure of the total bonding in any crystal lattice is given by the lattice energy. This is the enthalpy change associated with bringing together, in the case of salt, a mole of $Na^+(g)$ cations and a mole of $Cl^-(g)$ anions to form salt. This quantity cannot be measure directly, so it is derived indirectly through the use of other enthalpy changes which are known. The formation of salt can be approached from the constituent elements around the 'Born–Haber' cycle in two ways (Fig. 2.9) . The formation of salt from its component elements in their standard state is the enthalpy of formation and is a directly measurable quantity. Alternatively, the formation can be broken down into component

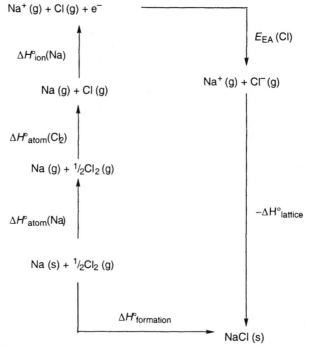

Fig. 2.9 Born–Haber cycle for NaCl.

processes, atomization, ionization, electron affinity, and the lattice energy. By Hess's law, the sum of these enthalpy changes of these component processes must add up to the enthalpy change of formation. So, provided the ionization enthalpies, atomization enthalpies, and electron affinity are known, the lattice energy may be calculated.

Lattice energies, in many cases, can be calculated. In cases where the experimentally derived lattice energies fail to match the calculated values based upon the compound being purely ionic, then the difference can often be ascribed to the compound not being a pure ionic structure; in other words, there is a covalent contribution to the structure. The form of the covalent contribution is a distortion of the electron clouds under the influence of the adjacent cations. There is a covalent component to the bonding in most ionic crystals, but it may be small.

2.3 Covalent bonding

In common salt, NaCl, chlorine gains an electron from sodium to form chloride and so achieve an octet. The situation for chlorine in molecules such as chlorine, Cl_2, is related, except that here a different mechanism is used by chlorine to achieve the octet structure. The chlorine atom has seven valence electrons and only requires one extra electron to attain the neon electronic structure. Instead of gaining an electron from an element such as sodium, it does this is by *sharing* an electron with another chlorine atom. In effect, the shared electrons (negatively charged) hold the positively charged nuclei

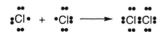

Lone pair: a pair of valence electrons not shared with another atom.

together through electrostatic attraction. For convenience, the two electrons connecting the chlorine atoms are normally represented by a line and so chlorine is Cl—Cl. Chlorine has six pairs of electrons that are *not* shared with other atoms. These pairs of electrons are *lone pairs*.

Oxygen has six valence electrons and so requires *two* electrons to attain the structure of neon, the noble gas. Rather than gaining two electrons by transfer from an atom such as magnesium, the oxygen atom can do this by sharing *two* electrons with a neighbouring oxygen atom. This places *two* shared pairs of electrons between the two oxygen nuclei. This introduces the concept of *bond order*. Two electrons located between two atoms are defined as forming a *single bond*. Two pairs of electrons between two atoms constitute two bonds, that is, a *double bond*. The bond order in dioxygen is 2 and dioxygen is written O=O. Each oxygen atom has two lone pairs.

Homonuclear diatomic molecules such as Cl_2, O_2, and N_2 are symmetrical and therefore there cannot be a charge imbalance between the two atoms. Diatomic molecules such as HCl are *asymmetric* and there is no reason, except by accident, for there *not* to be a charge imbalance between the two atoms. The power for atoms within a molecule to attract electrons to themselves is called electronegativity. An *electronegativity scale* (Chapters 1 and 5) is a set of numbers which ranks the ability of any given atom in a molecule to attract electrons from other atoms. Elements which attract electrons strongly are said to be very electronegative. Elements that do not do this well are weakly electronegative, or electropositive.

In a diatomic molecule such as H—Cl, the chlorine is more electronegative and attracts electron density away from the hydrogen. The consequence of this is that there is more electron density at the chlorine end of the molecule than at the hydrogen end. The bond is *polar*.

Two ways to denote that the H—Cl bond is polar.

The charge imbalance is an *ionic contribution* to the otherwise covalent H—Cl bond. Most bonds with different atoms at either end of the bond show a charge imbalance, and so a degree of ionic nature.

The procedure for writing down the Lewis structure for a polyatomic molecule starts with a calculation of the total number of valence electrons for all the atoms in a molecule, including any necessary compensation for charge if the molecule is ionic. The electrons are then arranged around the component atoms (arranged in the correct connectivity) so that, if possible, the octet rule (or duplet for hydrogen) is satisfied for each (Fig. 2.10). Note that the Lewis representation of ethene successfully accounts for the double bond. Atoms in molecules do not necessarily share all their electrons. Nitrogen in ammonia, NH_3, retains a lone pair of electrons for itself. Note that nitrogen still has an outer octet of electrons.

There are a very few exceptions to the octet rule for the second period elements. However, in BF_3, although all the fluorine atoms attain the octet configuration, boron has only six. The chemistry of molecules of this type is dominated by a tendency to react with a source of electrons, so attaining the octet configuration. The octet rule is frequently adhered to by main group elements below the second period, but there is scope for exceptions. So for the third period elements, there are a number of cases, such as PF_5, where the

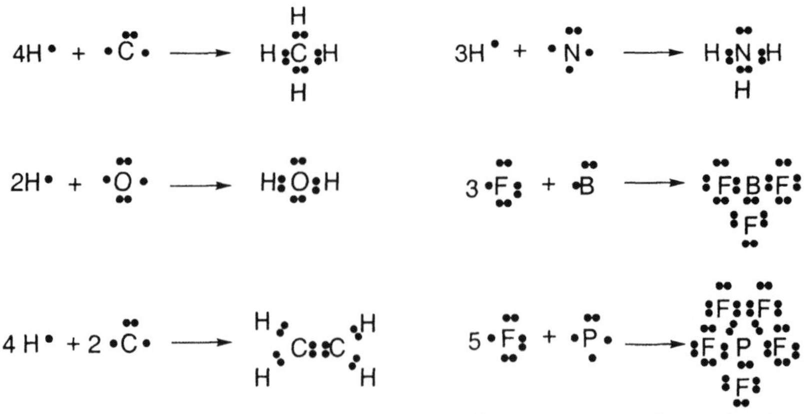

$$4H^{\bullet} + {}^{\bullet}\overset{\bullet\bullet}{\underset{\bullet\bullet}{C}}{}^{\bullet} \longrightarrow H \overset{\overset{\textstyle H}{\bullet\bullet}}{\underset{\underset{\textstyle H}{\bullet\bullet}}{\overset{\bullet\bullet}{C}}} H \qquad\qquad 3H^{\bullet} + {}^{\bullet}\overset{\bullet\bullet}{\underset{\bullet}{N}}{}^{\bullet} \longrightarrow H \overset{\overset{\textstyle H}{\bullet\bullet}}{\underset{\underset{\textstyle H}{}}{\overset{}{N}}} H$$

$$2H^{\bullet} + {}^{\bullet}\overset{\bullet\bullet}{\underset{\bullet\bullet}{O}}{}^{\bullet} \longrightarrow H \overset{\bullet\bullet}{\underset{\bullet\bullet}{O}} H \qquad\qquad 3 {}^{\bullet}\overset{\bullet\bullet}{\underset{\bullet\bullet}{F}}{}^{\bullet} + {}^{\bullet}B \longrightarrow$$

Fig. 2.10 and PF₃ and BF₃ Lewis structure diagrams

$$4H^{\bullet} + 2 {}^{\bullet}C^{\bullet} \longrightarrow \overset{\textstyle H}{\underset{\textstyle H}{}} C {::} C \overset{\textstyle H}{\underset{\textstyle H}{}} \qquad\qquad 5 {}^{\bullet}F {}^{\bullet} + {}^{\bullet}P^{\bullet} \longrightarrow PF_5\text{ structure}$$

Fig. 2.10 Some examples of Lewis structures used to represent covalent molecules.

electron count expands beyond eight. This is often attributed to participation in the bonding by *d* orbitals. There is space in the valence shell for more than eight electrons.

It is common to denote bonds by a line such as that in F—F. The line corresponds to two electron dots in Fig. 2.10 and is generally a more convenient representation than the dots. The two dots are retained for lone pairs, but lone pairs are often only denoted on the atom of interest in a molecule (Fig. 2.10). So in PF₃, only the phosphorus lone pair is generally written (:PF₃) and the fluorine lone pairs are omitted from the diagram.

One can elaborate on the covalent bonding in hydrogen as follows. The hydrogen atom has one electron (H'). In the molecule H₂, each H atom attains the electronic configuration of the next noble gas, in this case the helium duplet, by sharing the two electrons. This is represented as H:H, or more simply H—H. There are two electrons located between the two hydrogen nuclei and these two electrons form the bond holding the nuclei together.

There are a number of interactions (Fig. 2.11) to be considered. The force between two point charges varies with the square of the distance between them. The two protons are positively charged and repel each other. The two electrons are negatively charged and repel each other. However, this is more than offset by the four proton–electron attractions. These six interactions therefore represent the net bonding in the system.

In an orbital description of bonding between two hydrogen atoms, the two hydrogen atomic orbitals (1*s*) overlap to make an *orbital* which is distributed over both component atoms. The two electrons reside in the one orbital. The electron density aggregates between the two nuclei. This is shown schematically in Fig. 2.12 in a number of different styles, all of which are in common use.

In this model, bonds are regarded as *localized interactions involving two electrons and two atoms.* The localized bond approach implies that one orbital on one atom interacts with an orbital on a second atom to form a bond. Orbitals on other atoms in the molecule are not involved. A further approach (the molecular orbital description) involves delocalization of electrons in

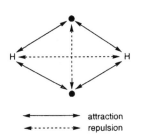

attraction
------ repulsion

Fig. 2.11 The interactions between the two hydrogen nuclei and the two electrons of H₂.

Fig. 2.12 A diagram (left) representing the electron density within H_2^+. A schematic representation of the electron cloud in H_2^+ is shown in the centre. A boundary representation of the H_2^+ molecular orbital containing the electron, together with the constituent atomic orbitals, is shown to the right.

molecular orbitals over the whole molecule, but this approach is beyond the scope of this text.

One way to describe these localized interactions is to use the *hybrid orbital* approach developed by Pauling. Hybridization, or *mixing of orbitals* on an atom, is a useful mathematical device. It is *not* something that atoms *do* in the sense that hybridization does not occur during chemical reactions, it is a convenient mathematical procedure which describes bonding in terms of conceptually simple two-centre, two-electron bonds. It gives a way of using modified atomic orbitals to overlap more strongly with atomic orbitals on other atoms and so to form more stable bonding orbitals.

Boron trihydride, BH_3, is trigonal planar. The three valence electrons in the boron atom are contained in a spherical s orbital (two electrons) and in a p orbital (one electron). It is not immediately clear how this arrangement of electrons can result in a trigonal molecule.

$$3H\cdot + \ddot{B}\cdot \longrightarrow H:B\overset{\cdot\cdot H}{\underset{\cdot H}{}}$$

The first conceptual requirement is to rearrange the electrons so that there are three orbitals each containing one electron. These three orbitals can then participate in shared electron bonds with the three hydrogen $1s$ orbitals. This can be achieved by first imagining that one of the electrons in the boron $2s$ orbital moves to an unoccupied boron p orbital (Eqn 2.1). This is often called promotion, since the energy level of the p orbital is above that of the s orbital. Of the three orbitals containing one electron, two are at 90° to each other and the other is non-directional. Unfortunately, these cannot be used directly for constructing three trigonal bonds (120° bond angles).

$$[\text{He}]2s^2 2p^1 \xrightarrow{\text{promotion}} [\text{He}]2s^1 2p_x^1 2p_y^1 \tag{2.1}$$

The shape of trigonal molecules is determined by three electron pairs. The most convenient way of generating acceptable trigonal orbitals on boron is to *hybridize* one s with two p orbitals (Eqn 2.2). Recall each of these contains one electron. By convention, the z-axis is taken to be perpendicular to the plane of the molecule. It makes sense to mix the $2p_x$ and $2p_y$ orbitals since the $2p_z$ orbital is directed out of the plane of the three bonds. These orbitals are called sp^2 hybrids. The three resulting orbitals are equivalent and each of the

The axis definition for BH_3.

empty p_z orbital

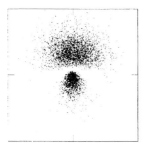

Fig. 2.13 Density representation of an sp^2 hybrid orbital.

hybrids contain one-third s character and two-thirds p character. One sp^2 hybrid is illustrated in Fig. 2.13. Each of these sp^2 hybrid orbitals contains one electron. The remaining $2p$ orbital ($2p_z$) is both empty and non-bonding. The sp^2 hybrid orbitals are now set up in the correct orientation to overlap with the three hydrogen atoms (Fig. 2.14) and so to form three shared electron bonds. Each of these resulting covalent bonds is called a σ bond. The electron density lies directly between the two connected nuclei.

$$[He]2s^1 2p_x^1 2p_y^1 \xrightarrow{\text{hybridization}} [He]\left(sp^2\right)^3 \qquad (2.2)$$

Systems based upon a tetrahedral geometry (Chapter 5) such as methane or SiH_4 require that the hybridization involves four orbitals. This means mixing the s and all three p valence orbitals into sp^3 hybrids. All four are equivalent, differing only in direction. Each contains one-quarter s character and three-quarters p character. In practice, they are very similar in appearance to sp^2 hybrid orbitals. The mathematics works out so that the orbitals are mutually orientated at angles of 109.5° — that is, along tetrahedral axes.

The ground state electronic structure of carbon is $[He]2s^2 2p^2$. Four unpaired electrons are achieved by conceptually promoting one of the electrons into the remaining empty p orbital (Eqn 2.3). Note that of the original four carbon orbitals, three are orientated at 90° and the fourth is non-directional. Without invoking hybridization to form sp^3 hybrids (Eqn 2.4) it would be difficult to construct four tetrahedral localized bonds. So in the hybrid view of bonding for methane there are four equivalent C—H bonds with bond angles of 109.5°.

$$[He]2s^2 2p^2 \xrightarrow{\text{promotion}} [He]2s^1 2p_x^1 2p_y^1 2p_z^1 \qquad (2.3)$$

$$[He]2s^1 2p_x^1 2p_y^1 2p_z^1 \xrightarrow{\text{hybridization}} [He]\left(sp^3\right)^4 \qquad (2.4)$$

Other hybridization schemes:
Trigonal bipyramidal — dsp^3
Octahedral — d^2sp^3
Square planar — dsp^2

Molecules such as PF_5 and SF_6, and most metal complexes, have more than four bonds and so require hybridization schemes that produce five, six, or more orbitals capable of binding to the peripheral atoms. This is achieved by involving d orbitals.

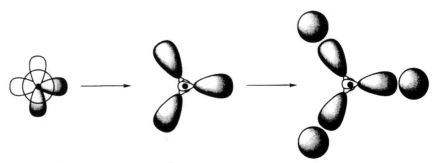

Unhybridized B sp^2 hybridized B   Orbital overlaps for BH₃

Fig. 2.14 The overlap of boron sp^2 hybrids with hydrogen $1s$ orbitals to form BH_3.

2.4   Oxidation state

Before proceeding further, it is necessary to introduce one further concept: the *oxidation state*. The oxidation state is a useful way to keep track of electrons and can be helpful when categorizing compounds.

Oxidation: removal of electrons

Reduction: gaining of electrons

Consider lithium fluoride, LiF. The formation of the salt involves the transfer of electrons from lithium to fluorine. The lithium is said to oxidize and the fluorine is said to reduce. Since a single electron is removed from each lithium atom, the oxidation number (or oxidation state) is +1. Since a single electron is added to each fluorine atom, the oxidation number (or oxidation state) is –1. Note the oxidation state has a sign.

$$Li \rightarrow Li^+ + e^-$$
$$F + e^- \rightarrow F^-$$

One can also assign oxidation numbers without considering formation processes. Most formulae, such as LiF, can be dissected in our minds into ions so as to ensure that all of the resulting ions possess a closed shell of electrons. For lithium fluoride, note that both the resulting ions Li^+ and F^- possess closed shell structures. The oxidation number is then defined as the charge possessed by each ion (+1 and –1 for LiF). Oxidation numbers are commonly written using Roman numerals, so the lithium in LiF is denoted Li(I) and the fluorine as F(–I). Aluminium oxide, Al_2O_3, is dissected as $2Al^{3+}$ and $3O^{2-}$ giving oxidation states of +3 and –2 for aluminium and oxygen.

In cases such as ClF where it is not possible to dissect the structure so that each ion possesses a closed shell structure, the electrons are assigned to the most electronegative atoms first. So ClF is dissected into Cl^+ and F^- and the oxidation number of chlorine is +1 and that of fluorine is –1. In this case after the dissection, the chlorine ion has only six valence electrons. It is far more usual for chlorine to adopt an oxidation state of –1 and so achieve an octet.

The above procedure is a common way to assign oxidation state and works well. However, the method can lead to the common misconception that the oxidation state corresponds to the electric charge on atoms in any particular compound. For salts such as LiF, perhaps regarding the structure as a lattice of Li^+ and F^- ions is not unreasonable, but consider CF_4. Here the oxidation states are assigned as +4 for carbon and –1 for each fluorine. However, tetrafluoromethane is covalent and while it is correct to assign carbon as C(IV) and the fluorine as F(–I), CF_4 is certainly not a lattice of C^{4+} and F^- ions. Uranium hexafluoride, UF_6, formally is made up from U(VI) and F(–I), but its nature is not an ionic lattice of U^{6+} and F^- ions. Instead, it is a covalent material, boiling point 56°C, which in the gas phase is used for the isotopic fractionation of uranium isotopes.

So, it is most important to realize that oxidation state is not a physical property. There is not a machine which functions as an oxidation state meter. Nevertheless, the oxidation state is a useful way to keep track of electrons and to categorize compounds. It is still in common use, although perhaps, in time, alternative general purpose categorization schemes might replace the oxidation number concept.

3   Hydrogen

3.1   The element

Hydrogen is by far the most abundant element in the universe and, indeed, all the other chemical elements were made from it. In many ways, hydrogen is unique and, although it is hardly found on earth as the free element, life here would be impossible without it (see Fig. 3.1). It accounts for virtually all of our energy, either directly in the form of sunlight, or indirectly as fossil fuels. Without hydrogen there would be no water to drink and DNA molecules would not form the double-helix structure that allows our genetic code to be copied and passed on to future generations.

Whilst the atoms of all other elements contain neutrons in their nuclei, the main isotope of hydrogen, ^1H, contains only a single proton. Hydrogen gas consists of diatomic molecules, H_2, with the atoms linked by a single covalent bond. The gas is conveniently prepared in the laboratory by adding a fairly reactive metal to dilute sulphuric or hydrochloric acid (Eqn 3.1).

$$Zn(s) + H_2SO_4(aq) \rightarrow ZnSO_4(aq) + H_2(g) \qquad (3.1)$$

In the UK, annual production of hydrogen gas is about $1 \times 10^9 \, m^3$ (about 100 000 tonnes). Some is formed during petroleum 'cracking'. Hydrogen is also obtained during the manufacture of chlorine and sodium hydroxide via electrolysis of sodium chloride solution. However, the main industrial source of hydrogen is the reaction between methane and steam at high temperature in the presence of a catalyst. The overall reaction is represented by Eqn 3.2. The carbon dioxide is removed by 'scrubbing' the product with potassium carbonate solution (Eqn 3.3). After use, the potassium carbonate is regenerated by heating with steam to reverse the reaction.

$$CH_4(g) + 2H_2O(g) \rightarrow 3H_2(g) + CO_2(g) \qquad (3.2)$$

$$K_2CO_3(aq) + H_2O(l) + CO_2(g) \rightarrow 2KHCO_3(aq) \qquad (3.3)$$

3.2   Binary compounds — the hydrides

According to their electron configurations, both members of the first period, hydrogen ($1s^1$) and helium ($1s^2$), are *s*-block elements, but in which groups should they be placed? Although helium has the same outer electron configuration as the metals of Group 2, common sense seems to dictate that it should be placed with the other unreactive noble gases in Group 18.

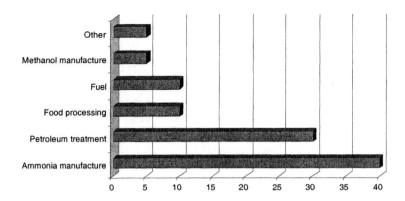

Fig. 3.1 Some of the most important industrial uses of hydrogen.

Hydrogen can form a positive ion like the metals of Group 1 but, since it is one electron short of a noble gas configuration, it can also form a negative ion and a single covalent bond, like the halogens of Group 17. Perhaps there is even a case for placing hydrogen with carbon in Group 14 as both elements have atoms with a half-filled outer shell of electrons and similar electronegativities. For these reasons, hydrogen and helium are sometimes in a separate block at the top of the periodic table, and are not included in the normal group classification. The periodic table (inside back cover) shows hydrogen as it is commonly placed. Other tables sometimes show hydrogen in both Groups 1 and 17. Table 3.1 compares some of the properties of hydrogen with those of the period 2 elements, lithium, carbon, and fluorine.

Hydrogen forms compounds with many other elements but the properties of these hydrides vary considerably. Some metals combine directly on heating with hydrogen to give ionic solids containing the hydride ion, H^-. Figure 3.2 shows that the formation of the hydride ion is energetically much less favourable than the fluoride ion, F^-. This means that only metals that ionize very easily can form ionic hydrides. This behaviour is, therefore, restricted to electropositive metals such as those in Group 1 and towards the bottom of Group 2. Even so, the product is energetically more stable than the separate elements as illustrated in the Born–Haber cycle for the formation of lithium hydride (Fig. 3.3). As might be expected, the ionic hydrides are

Table 3.1 A comparison of some properties of hydrogen, lithium, carbon and fluorine

Property	Hydrogen	Lithium (group 1)	Carbon (group 14)	Fluorine (group 17)
Electron configuration	$1s^1$	$[He]2s^1$	$[He]2s^22p^2$	$[He]2s^22p^5$
First ionization energy/ kJ mol^{-1}	+1310	+519	+1090	+1680
Electron affinity /kJ mol^{-1}	−67	−52	−120	−348
Electronegativity (Pauling)	2.1	1.0	2.5	4.0
Structure	diatomic molecules	metallic lattice	covalent macromolecule	diatomic molecules
Atomization energy/kJ mol^{-1}	+218	+161	+715	+79

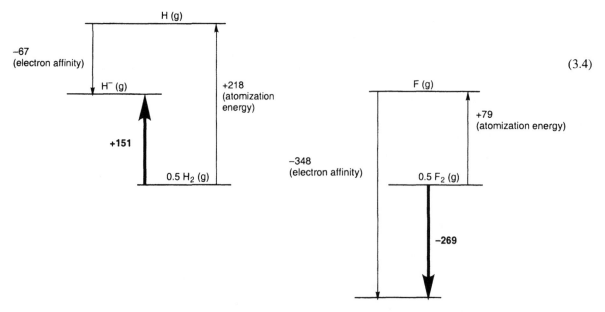

Fig. 3.2 Energetics of the formation of H⁻(g) and F⁻(g) (all values in kJ mol⁻¹).

generally rather unstable, reactive materials. They all react vigorously with water giving an alkaline solution and hydrogen gas (Eqn 3.4).

$$Li^+H^-(s) + H_2O(l) \rightarrow H_2(g) + Li^+(aq) + OH^-(aq)$$

The vast majority of simple hydrides are covalent, that is, where the hydrogen atom shares an electron from another atom to achieve the helium electronic configuration. The energy cycle drawn in Fig. 3.4 for ammonia, NH_3, typifies the process of forming a covalent hydride from its elements. A range of simple covalent hydrides formed by *p*-block elements is shown together with selected properties in Table 3.2.

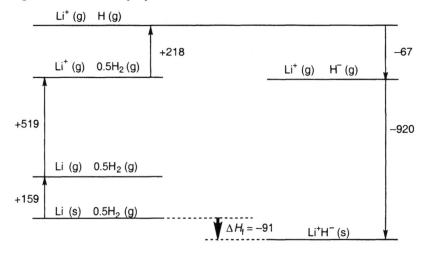

Fig. 3.3 Born–Haber cycle for the formation of lithium hydride from its elements (all values in kJ mol⁻¹).

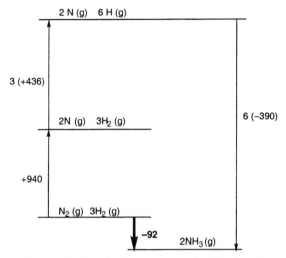

Fig. 3.4 Energy changes involved in forming ammonia, NH_3, from its elements (all values in kJ mol^{-1}).

The ΔH_f° values show that the energetic stability of the covalent hydrides tends to increase on passing from left to right across a period, but to decrease on passing down any group. The main reason for this is the change in the strength of the X—H bond as shown by the bond enthalpy figures in Table 3.2. The attraction of an atom for a shared pair of electrons increases across a period as nuclear charge increases and atomic radius falls, but decreases down a group as both atomic size and the number of inner screening electrons increase.

Although all of the hydrides shown in Table 3.2 are described as covalent, the bonds are polar to some extent. The degree of this polarity depends upon the electronegativity difference between the two elements concerned. In the case of the Group 14 hydrides, however, this bond polarity does not result in a polar molecule. In the tetrahedral XH_4 molecule, there is no overall dipole moment as the centre of the δ^+ charges coincides with the centre of the δ^- charge, for example in methane.

As they have no overall dipole, the only forces of attraction between XH_4 molecules are relatively weak van der Waals forces due to temporary dipoles. Since these forces result from mutual repulsion of the electron clouds of neighbouring molecules, they increase as the size of the molecule increases, thus accounting for the steady increase in the boiling point of the hydrides on passing down Group 14.

In the case of the hydrides of Groups 15, 16, and 17, bond polarity *does* result in a permanent molecular dipole (Fig. 3.5).

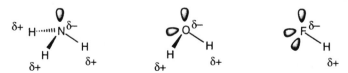

Fig. 3.5 Permanent dipoles in period 2 hydrides.

Table 3.2 Some common covalent hydrides of period 2. Electronegativity differences in brackets indicate that hydrogen is more electronegative than the *p*-block element, that is, the polarity of the bond is $X^{\delta+}$—$H^{\delta-}$

	Group 14	Group 15	Group 16	Group 17
Period 2	CH_4	NH_3	H_2O	HF
ΔH_f/kJ mol^{-1}	−74	−46	−286	−271
bond enthalpy/kJ mol^{-1}	+436	+431	+494	+574
boiling point/K	112	238	373	293
electronegativity difference (Pauling)	0.35	0.84	1.24	1.78
Period 3	SiH_4	PH_3	H_2S	HCl
ΔH_f/kJ mol^{-1}	34	-10	+20	−92
bond enthalpy/kJ mol^{-1}	+323	+323	+364	+432
boiling point/K	161	185	213	188
electronegativity difference (Pauling)	(0.3)	(0.01)	0.38	0.96
Period 4	GeH_4	AsH_3	H_2Se	HBr
ΔH_f/kJ mol^{-1}	+91	+66	+73	−36
bond enthalpy/kJ mol^{-1}	289	–	–	+363
boiling point/K	185	211	232	206
electronegativity difference (Pauling)	(0.19)	(0.02)	0.35	0.76
Period 5	SnH_4	SbH_3	H_2Te	HI
ΔH_f/kJ mol^{-1}	+163	+145	+100	+26
bond enthalpy/kJ mol^{-1}	253	–	–	+295
boiling point/K	221	256	269	238
electronegativity difference (Pauling)	(0.24)	(0.2)	(0.1)	0.46

In these cases, as well as attraction due to temporary dipoles, there is an extra attraction due to these permanent molecular dipoles. The positively charged hydrogen atom will attract a lone pair of electrons on the *p*-block atom of a neighbouring molecule (see Fig. 3.6 for the situation in hydrogen fluoride).

As a result of this extra permanent dipole attraction, in any period the boiling points of the Group 15, 16, and 17 hydrides are all higher than that of the non-polar Group 14 hydride. This effect is much greater in period 2 than in the lower periods (Fig. 3.7). Since the attraction resulting from the

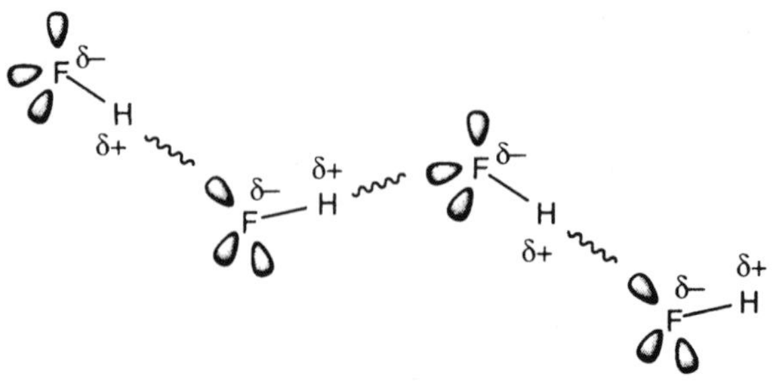

Fig. 3.6 Hydrogen bonding in hydrogen fluoride.

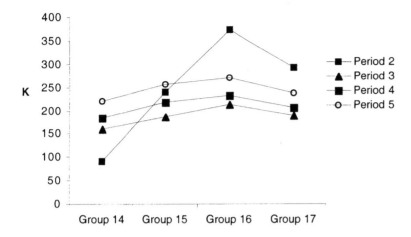

Fig. 3.7 Boiling point of covalent hydrides plotted by period.

presence of N—H, O—H, or F—H bonds is much stronger than other permanent dipoles, in these cases only the effect is referred to as *hydrogen bonding*. Typically such 'hydrogen bonds' are about 5–10% of the strength of a covalent bond and in the case of water this is enough to make it liquid at room temperature.

At first glance it might seem surprising that hydrogen fluoride has a lower boiling point than water, even though it contains the more polar H—F bond. Whilst the individual hydrogen bonds in HF are indeed stronger than those in water, the latter can form more hydrogen bonds. Each water molecule contains two $H^{\delta+}$ atoms and each $O^{\delta-}$ has two lone pairs and so may form up to four hydrogen bonds with its neighbours (Fig. 3.8).

On average both HF and NH_3 can form only two hydrogen bonds with neighbouring molecules and, as expected, hydrogen bonding between the less polar NH_3 molecules results in a lower boiling point for ammonia than hydrogen fluoride. The acid–base properties of covalent hydrides are covered in the next section.

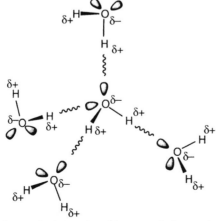

Fig. 3.8. Water can hydrogen bond to several other water molecules.

LIVERPOOL JOHN MOORES UNIVERSITY
LEARNING SERVICES

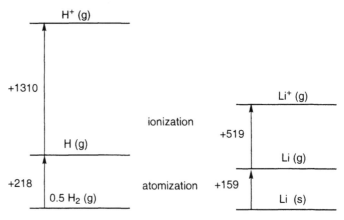

Fig. 3.9 Energetics of forming H⁻(g) and Li⁺(g) from the elements (all values in kJ mol⁻¹).

3.3 The hydrogen ion — acids and bases

As Fig. 3.9 shows, it takes over twice as much energy to form the free proton, H^+, from hydrogen gas than it does to form Li^+ from lithium. As a result, unlike the Group 1 metals, hydrogen forms no compounds in which it is present as a simple cation. In the last section, however, it was shown that hydrogen atoms covalently bonded to more electronegative atoms developed a partial positive charge, for example in water.

The $H^{\delta+}$ atoms attract a lone pair of electrons on neighbouring molecules. As well as forming a hydrogen bond, however, it is possible for the lone pair of electrons to form a coordinate bond to the hydrogen atom. If this happens, the original covalent bond must break as shown in Fig. 3.10. Of course, it is possible for these electron movements to reverse themselves but, to a small extent at least, water does self-ionize and contains small amounts of $OH^-(aq)$ and $H_3O^+(aq)$. The hydroxonium ion, H_3O^+, may be considered as a complex formed from the free H^+ ion and the water molecule which acts as an electron pair donor.

This self-ionization behaviour is not restricted to water and such self-ionization is also possible in hydrogen fluoride and liquid ammonia (Eqns 3.5 and 3.6).

$$2HF \rightleftharpoons H_2F^+ + F^- \tag{3.5}$$

$$2NH_3 \rightleftharpoons NH_4^+ + NH_2^- \tag{3.6}$$

$$K_c = \frac{[H_3O^+(aq)][OH^-(aq)]}{[H_2O(l)]^2} \tag{3.7}$$

Brønsted and Lowry defined an acid as a proton donor and a base as a proton acceptor. On this basis, the self-ionization of water may be regarded

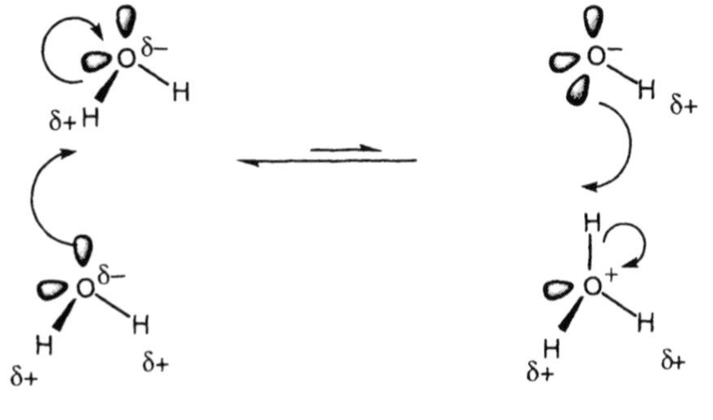

that is: 2H₂O OH⁻ + H₃O⁺

Fig. 3.10 The self-ionization of water.

as an acid–base reaction with one water molecule acting as an acid and the other as a base. The equilibrium constant for the self-ionization of water is written as in Eqn 3.7.

However, since very little of the water actually exists as ions, the concentration of water may be taken as constant. The H_3O^+ ion may also be written as $H^+(aq)$ (but *not* just H^+). This allows the use of Eqn 3.8, where K_w is referred to as the ionic product of water, as a simplified expression that applies to all dilute aqueous solutions.

$$K_w = [H^+(a)][OH^-(aq)]$$

Like any equilibrium constant, the value of K_w is temperature dependent, but at 25°C it has the value 1×10^{-14} M² (Eqn 3.9). In pure water at 25°C, the concentrations of $H^+(aq)$ and $OH^-(aq)$ must be equal (Eqn 3.10), leading to a value of 1×10^{-7} M for $H^+(aq)$. For convenience, a logarithmic scale is used for many of these quantities. In general, for any quantity X, a scale pX can be *defined* (Eqn 3.11). This leads to scales such as pH, pOH, and pK_w (Eqns 3.12–3.14). For pure water at 25°C, pH = pOH = 7 and pK_w = 14.

$$K_w = 1 \times 10^{-14} M^2 = [H^+(aq)][OH^-(aq)] = [H^+(aq)]^2 \qquad (3.9)$$

$$\text{So } \sqrt{K_w} = [H^+(aq)] = [OH^-(aq)] = 1 \times 10^{-7} \text{ M} \qquad (3.10)$$

$$pX = -\log_{10}X \qquad (3.11)$$

$$pH = -\log_{10}[H^+(aq)] \qquad (3.12)$$

$$pOH = -\log_{10}[OH^-(aq)] \qquad (3.13)$$

$$pK_w = -\log_{10}K_w. \qquad (3.14)$$

Pure water is neutral, regardless of temperature, since the two concentrations $[H^+(aq)]$ and $[OH^-(aq)]$ are equal. However, whilst K_w is fixed by temperature, the proportions of H^+(aq) and OH^-(aq) in an *aqueous solution* may be quite different. Any solution that has $[H^+(aq)] > [OH^-(aq)]$ is acidic, whilst in an alkaline solution $[H^+(aq)] < [OH^-(aq)]$. Hydrogen chloride dissolves in water to give an acidic solution, owing to its reaction with water (Eqn 3.15, or more simply, 3.16).

$$HCl(g) + H_2O(l) \rightarrow H_3O^+(aq) + Cl^-(aq) \qquad (3.15)$$

$$HCl(aq) \rightarrow H^+(aq) + Cl^-(aq) \qquad (3.16)$$

The HCl has acted as a Brønsted–Lowry acid by donating H^+ to H_2O acting as a Brønsted–Lowry base. As a result $[H^+(aq)]$ has *risen* whilst $[OH^-]$ has *fallen* to maintain the value of K_w. Hydrochloric acid is referred to as a strong acid since at equilibrium virtually all of it has ionized to donate protons.

Hydrogen fluoride also dissolves in water but gives a much less acidic solution (Eqn 3.17, or more simply, 3.18).

$$HF(l) + H_2O(l) \rightleftharpoons H_3O^+(aq) + F^-(aq) \qquad (3.17)$$

$$HF(aq) \rightleftharpoons H^+(aq) + F^-(aq) \qquad (3.18)$$

The extent to which HF acts as a proton donor, that is its acid strength, is given by the magnitude of its *acid dissociation constant*, K_a (Eqn 3.19).

$$K_a = \frac{[H^+(aq)].[F^-(aq)]}{[HF(aq)]} \qquad (3.19)$$

The mechanism for each of these reactions, illustrated in Fig. 3.11, shows that the first step involves the formation of a coordinate bond. Since the δ^+ charge on the H atom in HF is greater, this cannot explain why HCl is a stronger acid. However, the second step involves breaking the covalent bond in the hydrogen halide molecule. The bond energy values given in Table 3.2 show that the H—Cl bond is weaker than the H—F bond and this explains why the former ionizes more readily in aqueous solution.

Ammonia dissolves in water to give an alkaline solution (Eqn 3.20 or more

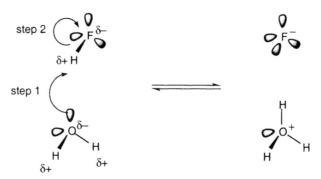

Fig. 3.11 The reaction of hydrogen fluoride with water.

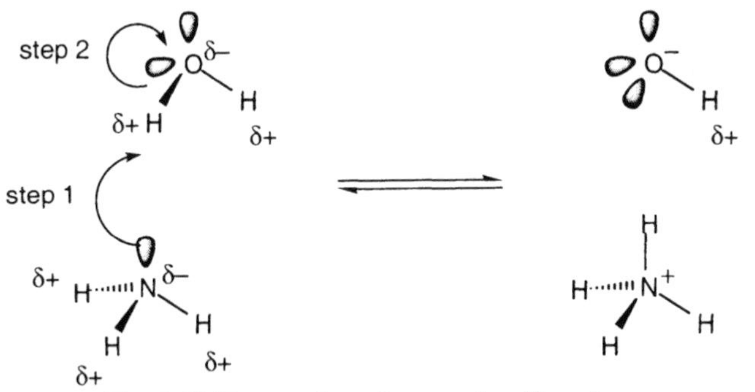

step 2

$\delta+$ H   H
$\delta+$ $\delta+$

step 1

$\delta+$ H······N$^{\delta-}$
H
H
$\delta+$ $\delta+$

$\Longrightarrow$

H
H······N^{+}
H
H

)

)

Fig. 3.12 The reaction of ammonia with water.

(3.22)

simply 3.21). Ammonia is a poorer proton donor than water so in this case acts as a Brønsted–Lowry base. The base strength of ammonia, that is the extent to which it accepts protons from water, is given by its *base dissociation constant* K_b (Eqn 3.22). The mechanism for the reaction of ammonia with water, is shown in Fig. 3.12.

$$NH_3(g) + H_2O(l) \rightleftharpoons NH_4^+(aq) + OH^-(aq)$$

$$NH_3(aq) \rightleftharpoons NH_4^+(aq) + OH^-(aq)$$

$$K_b = \frac{[NH_4^+(aq)][OH^-(aq)]}{[NH_3(aq)]}$$

Phosphine, PH_3, has a similar structure to NH_3, but has virtually no basic character. This cannot be because of step 2 (Fig. 3.12) which is identical in each case. However, the first step requires the donation of a lone pair of electrons from the Group 15 hydride to a hydrogen atom on the water molecule. Table 3.2 shows that the electronegativities of P and H are virtually the same so the P atoms in PH_3 have negligible partial charge and so cannot donate a lone pair of electrons as readily to the water molecule.

3.4 Sulphuric acid

Sulphuric acid, H_2SO_4 (Fig. 3.13), is perhaps the most important industrial chemical. It is made in vast quantities around the world. By 1980 about 5 megatonnes of sulphuric acid were being made annually in the UK, and about eight times that amount in the USA. It was made in Europe in the sixteenth century.

Sulphuric acid when anhydrous is an oily, dense, viscous liquid. It solidifies at about 11°C. It is miscible with water and the dissolution process is highly exothermic. Addition of water to the acid is dangerous since the rapid evolution of heat results in spattering. When it is necessary to mix sulphuric acid and water, the acid should be added to water, carefully and with stirring.

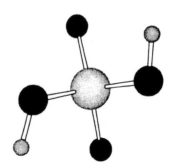

Fig. 3.13 Gas-phase structure of sulphuric acid.

The HSO_4^- ion is called bisulphate or hydrogensulphate.

The anhydrous acid is an electrical conductor. This is because of self-ionization (Eqn 3.23).

$$2H_2SO_4 \rightleftharpoons H_3SO_4^+ + HSO_4^- \qquad (3.23)$$

The self-ionization constant given by $[H_3SO_4][HSO_4^-]$ at 25°C is about 3×10^{-4}, which compares with the equivalent value of 10^{-14} for water.

Many metals dissolve in sulphuric acid. So addition of iron to sulphuric acid (originally known as oil of vitriol) results in iron(II) sulphate, $FeSO_4$. The sulphate ion, SO_4^{2-}, is tetrahedral. In addition to existing as a free tetrahedral species it also binds to *d*-block metals as a monodentate or bidentate ligand (Chapter 6).

Modern routes to sulphuric acid are largely via sulphur dioxide, SO_2. This is made by controlled oxidation of sulphur or by the roasting of sulphide minerals. Further oxidation of SO_2 in the presence of a catalyst results in SO_3, sulphur trioxide, which under controlled conditions is reacted with water to form sulphuric acid. The whole process of sulphuric acid production is highly exothermic and the effective utilization of the evolved energy is important in keeping the cost of the acid down. No process is ever 100% efficient, and some SO_2 does escape from the plant. Environmental controls limit the escape but considerable amounts of SO_2 are released into the atmosphere. There are many other far more serious sources of SO_2 in the atmosphere, in particular coal- and oil-fired power stations, oil refineries, and copper smelters. Sulphur dioxide in the atmosphere damages plant life at concentrations considerably lower (1–2 ppm) than actual permitted atmospheric levels and is one of the most important causes of acid rain.

3.5 Nitric acid

Nitric acid, HNO_3 (Fig. 3.14), is made largely by the catalytic oxidation of ammonia. Under the conditions, NO is formed and is further oxidized to nitrogen dioxide before treatment of the resulting NO_2 with water to form nitric acid (Eqns 3.24 – 3.26). Nitric acid turns brown on standing. This is because of decomposition in daylight, and the resulting formation of NO_2 (Eqn 3.27). Nitric acid self-ionizes very readily (Eqn 3.28) to form nitrate.

$$4NH_3 + 5O_2 \rightarrow 4NO + 6H_2O(l) \qquad (3.24)$$

$$2NO + O_2 \rightarrow 2NO_2 \qquad (3.25)$$

$$3NO_2 + H_2O \rightarrow 2HNO_3 + NO \qquad (3.26)$$

$$4HNO_3 \rightleftharpoons 4NO_2 + 2H_2O + O_2 \qquad (3.27)$$

$$2HNO_3 \rightleftharpoons [H_2NO_3]^+ + NO_3^- \qquad (3.28)$$

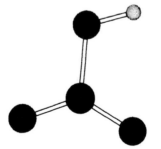

Fig. 3.14 Gas-phase structure of nitric acid.

4 *s*-Block elements

Elements with atoms whose last electron occupies an *s* sublevel orbital are *s*-block elements. In the standard periodic table these elements occupy the two columns farthest to the left. By this definition, the two columns would be headed by hydrogen and helium. However, in the *standard periodic table* helium is moved to head Group 18, because in common with other Group 18 elements, it is an unreactive gas.

In other formats of the periodic table based more strictly upon electronic configurations, helium is in Group 2.

Disregarding hydrogen (see Chapter 3) and helium (see Chapter 5), all the elements with their outermost electrons in an *s* sublevel are metals. Both columns in the *s*-block are referred to by special names. The elements of Group 1 are known as the *alkali metals*, whereas those of Group 2 are the *alkaline earth metals*. Some key data for the elements are shown in Table 4.1.

Table 4.1 Some data for the *s*-block elements

Alkali metals	Lithium	Sodium	Potassium	Rubidium	Caesium	Francium
Electron configuration	[He] $2s^1$	[Ne] $3s^1$	[Ar] $4s^1$	[Kr] $5s^1$	[Xe] $6s^1$	[Rn] $7s^1$
Metallic radius/ nm	0.156	0.191	0.235	0.248	0.267	0.270
Ionic radius M$^+$/ nm	0.060	0.095	0.133	0.148	0.169	0.176
1st ionization energy/ kJ mol^{-1}	519	494	418	402	376	381
2nd ionization energy/ kJ mol^{-1}	7300	4560	3070	2650	2420	
3rd ionization energy/ kJ mol^{-1}	11800	6940	4600	3850	3300	
Atomization energy/ kJ mol^{-1}	159	107	89	81	76	
Melting point/K	454	371	337	312	302	
Boiling point/K	1615	1156	1032	961	941	
Pauling electronegativity	0.98	0.93	0.82	0.82	0.79	0.7
E^o/ V for M$^+$(aq) $\rightarrow$ M(s)	−3.04	−2.71	−2.92	−2.92	−2.92	

Alkaline earth metals	Beryllium	Magnesium	Calcium	Strontium	Barium	Radium
Electron configuration	[He] $2s^2$	[Ne] $3s^2$	[Ar] $4s^2$	[Kr] $5s^2$	[Xe] $6s^2$	[Rn] $7s^2$
Metallic radius/ nm	0.112	0.160	0.197	0.215	0.222	0.220
Ionic radius M$^+$/ nm	0.031	0.065	0.099	0.113	0.135	0.140
1st ionization energy/ kJ mol^{-1}	900	736	590	549	502	510
2nd ionization energy/ kJ mol^{-1}	1760	1450	1150	1060	966	
3rd ionization energy/ kJ mol^{-1}	14800	7740	4940	4120	3390	
Atomization energy/ kJ mol^{-1}	324	146	178	164	182	159
Melting point/K	1560	923	1115	1050	1000	973
Boiling point/K	2742	1363	1757	1655	2143	2010
Pauling electronegativity	1.57	1.31	1.00	0.95	0.89	0.9
E^o/ V for M^{2+}(aq) $\rightarrow$ M$^+$(s)	−1.85	−2.38	−2.87	−2.89	−2.90	−2.92

4.1 Uses

Many foods contain compounds of these elements and many of these 'minerals' are essential for health. For example, calcium salts are needed, especially in infants, for bone formation. The distribution of sodium and potassium ions helps to regulate the water content of living cells. However, excessive consumption of sodium, for example by using too much salt on food, can lead to dangerously high blood pressure. It has been known since the early 1900s that lithium compounds exert a pharmaceutical effect. Doses of lithium carbonate, Li_2CO_3, may assist in the treatment of manic-depressive disorders. It is not clear how this happens but possibly the lithium ions in the body affect the sodium/potassium balance between cells and their surroundings.

Despite their high reactivity, the metals of the *s*-block have several commercially important uses. Sodium and magnesium are the most widely used but small quantities of the other metals are needed for special applications. Over 100 000 tonnes of sodium are currently used annually, alloyed with lead, in the manufacture of tetraethyllead, $Pb(C_2H_5)_4$, an 'anti-knock' additive for petrol. Owing to environmental concerns over lead pollution, however, the demand for sodium for this purpose is declining and likely to continue to decline.

$$Pb/4Na + 4C_2H_5Cl \rightarrow Pb(C_2H_5)_4 + 4NaCl \qquad (4.1)$$

The excellent thermal conductivity and low melting point of the Group 1 metals make them very efficient coolants, although special methods of handling are required given their high reactivity. 'Fast' nuclear reactors in France use circulating molten sodium to transfer the heat produced in their cores to steam boilers.

Magnesium is widely used in low-density alloys, which have excellent strength and heat resistant properties. These are used by the aircraft and aerospace industries and provide the fuel cans in 'magnox' nuclear reactors. Lithium is a constituent of specialist alloys and long-lasting lightweight batteries that are often found in computers, cameras, and other technological equipment.

Many compounds of *s*-block elements are also of industrial importance and large quantities of salt (sodium chloride) and limestone (calcium carbonate) are used directly and in the manufacture of other materials.

4.2 Occurrence and extraction

Although sodium, potassium, magnesium, and calcium are quite abundant in nature, they are never found in the free state as the metals are too reactive.

Since many *s*-block compounds, especially those cf Group 1, are appreciably water-soluble, they tend to be washed out into the sea. On average, seawater contains about 3.5% by mass of dissolved salts with a typical composition shown in Table 4.2. The relative abundances of sodium and potassium in the earth's crust are comparable, unlike those in seawater

Table 4.2 Abundances of Group 1 and 2 elements in seawater

Cations	% by mass	Anions	% by mass
Li^+	0.000018	F^-	0.0000130
Na^+	1.105	Cl^-	1.987
K^+	0.0416	Br^-	0.00673
Rb^+	0.000012	I^-	0.0000060
Cs^+	0.00000005	SO_4^{2-}	0.271
Be^{2+}	0.00000000006	CO_3^{2-}	0.014
Mg^{2+}	0.133		
Ca^{2+}	0.0422		
Sr^{2+}	0.081		
Ba^{2+}	0.000003		

where sodium predominates strongly. This is associated with the relative solubilities of sodium and potassium salts. Sodium salts are more soluble, and so are washed more easily into the sea.

Before the metals can be extracted from the sea, the seawater must be evaporated off to leave solid material. This is expensive in energy terms and it is often commercially preferable to use existing deposits of minerals such as rock salt (NaCl) which are formed by natural evaporation of ancient seas. Some of the main sources of these elements are shown in Table 4.3.

The reactive *s*-block metals are usually extracted by electrolysis of their molten chlorides. Sodium, for example, is extracted using the Downs cell illustrated in Fig. 4.1. To be electrolysed, liquid sodium chloride is required. Sodium chloride has a high melting point and it would be expensive in terms of energy to liquefy it. However, a mixture of sodium chloride (40%) and calcium chloride (60%) has a much lower melting point than pure sodium chloride and so helps to reduce energy costs. The reactions given by Eqns 4.1 and 4.2 occur in the cell at 580°C.

At the steel cathode: $Na^+ + e^- \rightarrow Na$ (also some $Ca^{2+} + 2e^- \rightarrow Ca$)   (4.1)

At the carbon anode: $2Cl^- \rightarrow Cl_2 + 2e^-$ (4.2)

Table 4.3. Group 1 and 2 element minerals

Element	Common sources
Lithium	Spodumene ($LiAlSi_2O_6$)
Sodium	Rocksalt (NaCl), trona (Na_2CO_3), saltpetre ($NaNO_3$)
Potassium	Sylvite (KCl), carnallite ($KCl\,MgCl_2\,6H_2O$)
Rubidium	Impurities in the lithium mineral lepidolite
Beryllium	Beryl ($Be_3Al_2Si_2O_6$)
Magnesium	Magnesite ($MgCO_3$); dolomite ($MgCO_3/CaCO_3$)
Calcium	Limstone ($CaCO_3$); gypsum ($CaSO_4 2H_2O$); fluorite (CaF_2)
Strontium	Celestite ($SrSO_4$); stronianite ($SrCO_3$)
Barium	Barite ($BaSO_4$)

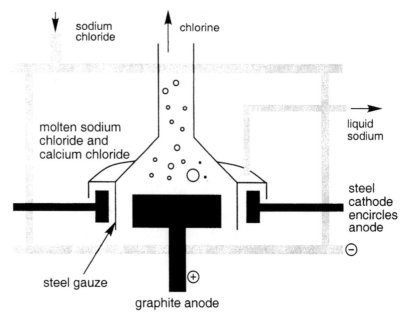

Fig. 4.1 Diagrammatic representation of the Downs cell.

Most of the calcium in the product crystallizes out on cooling and may be separated from the molten sodium. The solid calcium is returned to the electrolysis cell and reacts with sodium chloride to regenerate calcium chloride and to make new sodium (Eqn 4.3).

$$Ca + 2NaCl \rightarrow CaCl_2 + 2Na \tag{4.3}$$

Potassium metal cannot be made in this way because of its high solubility in the melt and because of its high volatility. Instead, reduction of molten potassium chloride, KCl, with metallic sodium at 850°C results in formation of metallic potassium in an equilibrium process. As potassium is more volatile it distils off more easily. This displaces the equilibrium, allowing the reaction to continue. Analogous processes are used to make metallic rubidium and caesium.

4.3 Structure of the elements

Atomization energy is defined as the enthalpy change on producing one mole of gaseous atoms from the element in its most stable state at 25°C and 1 atm pressure. For the s-block metals, this is represented by Eqn 4.4.

$$M(s) \rightarrow M(g) \tag{4.4}$$

Atomization energy and boiling point (Table 4.1) are *both* measures of the energy required to completely overcome the attractive forces holding the particles together in the metallic lattice. As Fig. 4.2 shows, there is a good correlation between these properties for the s block elements. In general, the

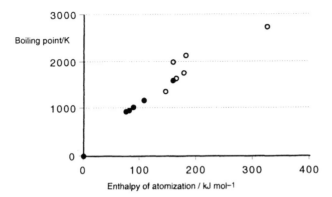

Fig. 4.2 The correlation between boiling point and atomization energy for the *s*-block elements. Solid circles are Group 1 elements, hollow circles are Group 2.

variation in boiling point is mirrored by the trend in melting point for these elements.

Metals consist of regular arrays of atoms. In the metallic crystal lattice, the outer electrons of the metal atoms interact with adjacent metal atoms. The interaction is an attraction between these electrons, which are negatively charged, and adjacent nuclei, which are positively charged. It is expressed as the interaction of the outer electron orbitals with those of the surrounding atoms. One way to view this is as the delocalization of outer electrons as a mobile 'sea' in which metal cations are embedded (Fig. 4.3).

Metallic bonding may be considered as the attraction between the nuclei of the metal ions and the delocalized electron 'sea'. The strength of attraction increases with density of the electron 'sea' and nuclear charge but decreases with increasing ionic radius and number of inner screening electrons.

The atomization energy, boiling point, and melting point increase on passing across any period from Group 1 to Group 2 (Table 4.1), but generally decrease on passing down either group. Deviations from this simple pattern in Group 2 are explained by differences in the way the metal atoms pack together in their metallic lattice. All Group 1 elements pack in a body-centred cubic lattice. However, in Group 2, beryllium and magnesium are hexagonal close packed, calcium and strontium are cubic close packed, and barium is

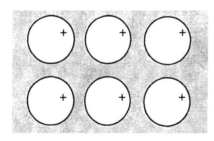
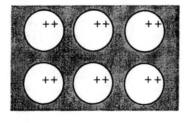

Group 1   Group 2

Fig. 4.3 The depth of shading in these metallic structures indicates the relative density of the delocalized electrons.

body-centred cubic.

4.4   Periodic trends down the *s*-block elements

Table 4.1 summarizes some of the main atomic and physical properties of the *s*-block elements. Atomic size, given here by the metallic radius, is influenced by three main factors:

- the *charge on the nucleus*. As this increases, the nucleus draws the electrons closer, so tending to reduce the size of the atom.
- the *number of shells containing electrons*. Starting an extra shell increases the atomic radius.
- the *number of inner electrons* that help to *screen* the outer electrons from the attraction of the nucleus. As this screening increases, the atom will increase in size. This effect is related to the number of electron shells in the atom.

On going across a period the only change is an increase in nuclear charge, so the Group 2 atom is always smaller than the Group 1 atom (Fig. 4.4). This is because of the increased ability of the nucleus to attract electrons. On passing down either group, the effect of increasing nuclear charge is outweighed by the addition of an extra occupied shell and the consequent increase in the number of inner screening electrons, resulting in an increase in atomic size.

When they react, the elements form cations in which the electrons in the outer shell have been removed. As the nuclear charge remains the same, whilst the number of electron shells and the number of screening electrons fall, the cations of the *s*-block are smaller than their parent atoms. The variation in atomic and ionic radius is shown graphically in Fig. 4.4. In general, cations are always smaller than the parent atoms.

Many other properties of the *s*-block elements are influenced by the same factors that determine atomic size. The first ionization energy (Fig. 4.5) is the energy change for removal of an electron from the gaseous atom (Eqn 4.5). Values are usually determined from optical spectra. The attraction of the nucleus for the most loosely bound electron must be overcome. This

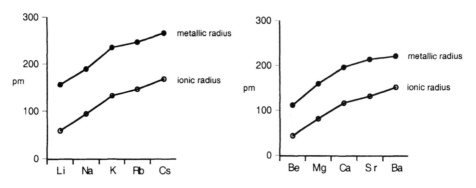

Fig. 4.4 Metallic and ionic radii for the Group 1 and 2 elements.

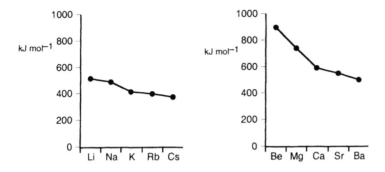

Fig. 4.5 First ionization energies for the Group 1 and 2 elements.

increases as nuclear charge rises but decreases as both the atomic size and number of inner screening electrons increase. Consequently, the first ionization energy of the alkaline earth metal (Group 2) is greater than that of the alkali metal (Group 1) in the same period, but the values decrease on passing down either group.

$$M(s) \rightarrow M^+(g) + e^- \tag{4.5}$$

Electronegativity is a measure of the attraction of an atom for the electrons in a covalent bond when that atom is in a molecule. In general, electronegativity decreases down the periodic table and increases from left to right across the table as shown in Fig. 4.6. Ionization energy is also a measure of the tendency to attract electrons, albeit in atoms rather than molecules. As shown in Fig. 4.7, there is a very good correlation between electronegativity and first ionization energy in the *s* block.

4.5 Standard electrode potential, *E°*

The standard electrode potential measures the tendency of an element to ionize in solution under standard conditions, i.e. 1 atmosphere pressure, 25°C, and 1 M aqueous concentrations. Consider Eqns 4.6 and 4.7.

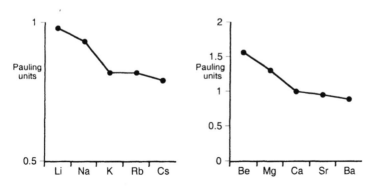

Fig. 4.6 Group trends in electronegativity of the Group 1 and 2 elements.

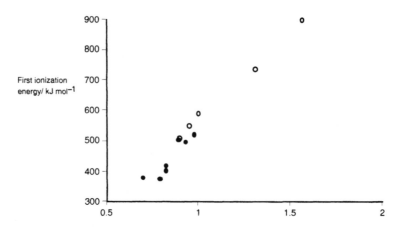

Fig. 4.7 The correlation between electronegativity and the first ionization energy. Solid circles are Group 1 elements and hollow circles Group 2 elements.

For the alkali metals: $M^+(aq) + e^- \rightleftharpoons M(s)$ (4.6)

For the alkaline earths: $M^{2+}(aq) + 2e^- \rightleftharpoons M(s)$ (4.7)

The more negative the value of $E°$, the further the equilibria in these equations lies over to the left, that is, in favour of the ions. Since the first ionization energy drops on passing down each group, $E°$ values should also steadily become more negative. This holds true for Group 2 (Table 4.1). However, Table 4.1 also shows that lithium at the top of Group 1 has the most negative $E°$ value of all.

The explanation for this lies in the energetics of forming aqueous ions from the metal. Apart from the first ionization energy, two other enthalpy changes are involved: atomization and hydration (for Group 1 elements: Eqns 4.8–4.10). As the data in Table 4.4 show, the energy released on hydration of the small lithium ion makes the overall enthalpy change for forming $Li^+(aq)$ from the metal more favourable than for any of the other Group 1 metals.

Atomization: $M(s) \rightarrow M(g)$ (4.8)

Ionization: $M(s) \rightarrow M^+(g) + e^-$ (4.9)

Hydration: $M^+(g) \rightarrow M^+(aq)$ (4.10)

4.6 Typical reactions

The chemistry of the elements is dominated by a tendency to lose electrons so forming M^+ in the case of the Group 1 elements or M^{2+} in the case of the

Table 4.4 Apart from the last column, all values in kJ mol⁻¹

Element	Atomization energy: $M(s) \rightarrow M(g)$	1st ionization energy: $M(g) \rightarrow M^+(g)$	Hydration energy: $M^+(g) \rightarrow M^+(aq)$	Overall enthalpy change: $M(s) \rightarrow M^+(aq)$	E^o/volts
Li	+159	+519	−519	+159	−3.04
Na	+107	+494	−406	+195	−2.71
K	+89	+418	−322	+185	−2.92
Rb	+81	+402	−301	+182	−2.92
Cs	+76	+376	−276	+176	−2.92

Group 2 elements. All the *s*-block metals react with electronegative non-metals, such as the halogens, to form ionic compounds. With chlorine, for example, the alkali metals form crystalline solids of general formula MCl. The enthalpy changes involved in forming NaCl(s) from its elements are shown in the Born–Haber cycle in Fig. 4.8.

The main 'driving force' for this reaction is the high lattice energy of sodium chloride, that is, the energy released on forming the solid ionic lattice from the separate gaseous ions. This more than compensates for the energy required to form the ions from the elements.

The alkaline earth metals also react with chlorine but form solid ionic compounds of general formula MCl₂. The Born–Haber cycle for the formation of MgCl₂ is shown in Fig. 4.9. Again, the stability of this compound is largely due to the lattice energy. The lattice energy is larger for MgCl₂ than NaCl as the Mg²⁺ ions attract Cl⁻ ions more strongly than Na⁺ ions.

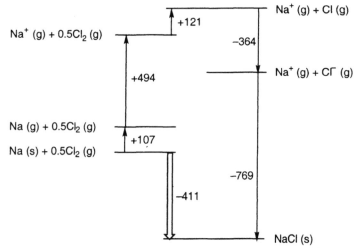

Fig. 4.8 Born–Haber cycle for NaCl(s) (all values in kJ mol⁻¹).

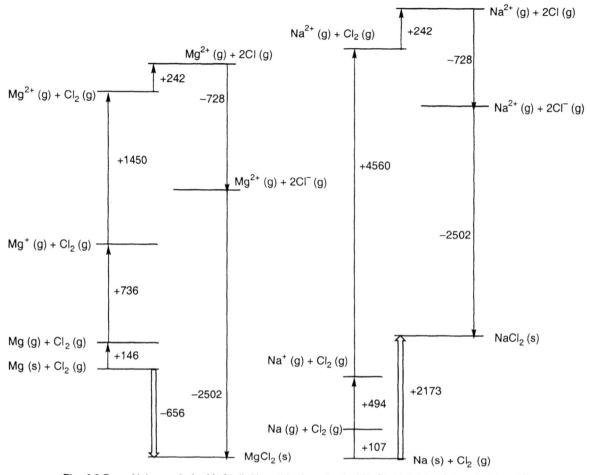

Fig. 4.9 Born–Haber cycle for MgCl$_2$ (left) and the hypothetical NaCl$_2$ (right) (all values in kJ mol^{-1}).

Why, then, doesn't sodium react with chlorine to form NaCl$_2$? If it is assumed crudely that the lattice enthalpy of this compound would be the same as that of MgCl$_2$, then the Born–Haber cycle for its formation is shown in Fig. 4.9. This shows that NaCl$_2$ would be less stable than its component elements. This is because of the very high value of the second ionization energy of sodium (Table 4.1). This involves removal of an electron from the inner $2p$ sublevel, a level that has much greater attraction to the nucleus than the valence $3s$ level. The s-block metals, therefore, form ionic compounds in which their atoms lose only the outer s electrons, as removal of any further electrons is energetically unfavourable.

Although all s-block metals react in a similar way with chlorine, different types of compounds may be produced on reaction with oxygen. Under some conditions, as well as forming oxide ions, O^{2-}, oxygen *molecules* may accept one or two electrons to form superoxide, O$_2^-$, and peroxide, O$_2^{2-}$, ions respectively (Fig. 4.10). When an excess of the metal is used, the simple oxide is always formed (Eqns 4.11 and 4.12).

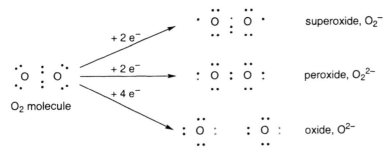

Fig. 4.10 The formation of superoxide, peroxide, and oxide from oxygen molecules.

$$\text{Group 1:} \quad 4M(s) + O_2(g) \rightarrow 2M_2O(s) \qquad (4.11)$$

$$\text{Group 2:} \quad 2M(s) + O_2(g) \rightarrow 2MO(s) \qquad (4.12)$$

The main products obtained on heating the metals in an excess of oxygen at atmospheric pressure are:

$Li \rightarrow Li_2O$	(oxide)	$Be \rightarrow BeO$	(oxide)
$Na \rightarrow Na_2O_2$	(peroxide)	$Mg \rightarrow MgO$	(oxide)
$K \rightarrow KO_2$	(superoxide)	$Ca \rightarrow CaO_2$	(peroxide)
$Rb \rightarrow RbO_2$	(superoxide)	$Sr \rightarrow SrO_2$	(peroxide)
$Cs \rightarrow CsO_2$	(superoxide)	$Ba \rightarrow Ba(O_2)_2$	(superoxide)

The ability of an *s*-block metal to form a peroxide or superoxide seems to depend upon the radius of its cation (Table 4.1). Only the largest cations form stable superoxides, while slightly smaller metal ions form peroxides, and cations smaller than Na^+ (95 pm) form only normal oxides. The reason is the weakness of the bond joining the oxygen atoms in O_2^- and especially in O_2^{2-}. If the cation's nucleus attracts electrons strongly enough, these bonds break, releasing oxygen gas and forming the normal oxide (Fig. 4.11).

The attraction of the cation for electrons decreases on passing down the *s*-block as the increase in ionic radius and the number of inner screening electrons outweigh the effect of increasing nuclear charge. Since the cations of Group 2 are smaller than their Group 1 counterparts, these elements are less likely to form peroxides and superoxides.

Ionic peroxides and superoxides are used to replace oxygen and remove carbon dioxide in submarines and manned spacecraft where it is impractical to carry sufficient supplies of 'fresh' air (Eqns 4.13–4.14).

electron drift

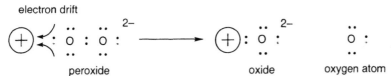

peroxide   oxide   oxygen atom

Fig. 4.11 Schematic representation of peroxide bond rupture.

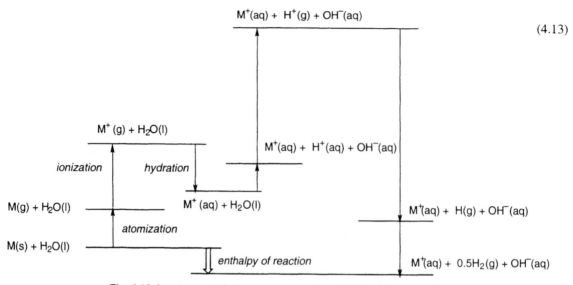

Fig. 4.12 An energy cycle for the reaction of an alkali metal with water.

$$O_2^{2-} + H_2O \rightarrow 2OH^- + 0.5O_2$$

followed by $2OH^- + CO_2 \rightarrow H_2O + CO_3^{2-}$ (4.14)

All the alkali metals react with cold water to form hydrogen gas and the aqueous metal hydroxide (Eqn 4.15). The violence of this reaction increases satisfyingly down the group.

$$M(s) + H_2O(l) \rightarrow M^+(aq) + OH^-(aq) + 0.5H_2(g) \quad (4.15)$$

An energy cycle for this reaction is shown in Fig. 4.12. Only three of the enthalpy changes involve the metal: atomization, ionization, and hydration of $M^+(g)$ (for values see Table 4.5). The remaining enthalpy changes involve only water and its products (Fig. 4.13).

The reaction with water becomes more exothermic and more vigorous on passing down Group 1. From calcium downwards the alkaline earth metals also react with cold water, although less vigorously than their Group 1 counterparts (Eqn 4.16).

$$M(s) + 2H_2O(l) \rightarrow M^{2+}(aq) + 2OH^-(aq) + H_2(g) \quad (4.16)$$

	$\Delta H/\text{kJ mol}^{-1}$
$H_2O(l) \rightarrow H^+(aq) + OH^-(aq)$	+57
$H^+(aq) \rightarrow H^+(g)$	+1091
$H^+(g) + e^- \rightarrow H(g)$	−1310
$H(g) \rightarrow 0.5H_2(g)$	−218
$H_2O(l) + e^- \rightarrow OH^-(aq) + 0.5H_2(g)$	−380

Fig. 4.13 The reduction of water by an electron.

Table 4.5 Atomization, ionization, and hydration enthalpies of $M^+(g)$

ΔH/kJ mol^{-1}	Li	Na	K	Rb	Cs
Atomization, M	+159	+107	+89	+81	+76
Ionization, M	+519	+494	+418	+402	+376
Hydration, M^+	−519	−406	−322	−301	−276
(Water)	−380	−380	−380	−380	−380
Overall	−221	−185	−195	−203	−204

4.7 Compounds

Most of the compounds of the alkali and alkaline earth metals have typical ionic properties, So, they are crystalline solids with high melting points that conduct electricity when molten or in solution. As with the elements, however, some variations in the properties of the compounds are apparent on passing down each group. The water solubility data for the chlorides, hydroxides, sulphates, and carbonates are summarized in Table 4.6.

Although a full treatment of this topic involves free energy, the table shows that, within any particular series of compounds, molar solubility generally increases as the enthalpy of solution becomes more exothermic. The overall process of dissolution can be considered as taking place in stages. Energy must be supplied to separate the ions in the lattice but is then released when the individual ions are hydrated. As an example, compare sodium hydroxide with magnesium hydroxide. The enthalpy changes involved for each compound are shown in Figs 4.15 and 4.16.

Not only is the enthalpy of solution less favourable for magnesium hydroxide but also the higher lattice energy presents a bigger activation energy barrier to dissolving. Not surprisingly, sodium hydroxide is readily soluble in water, whereas magnesium hydroxide is only slightly soluble.

Table 4.6 Molar concentrations of saturated aqueous solutions of some s-block compounds at 293 K. Molar enthalpies of solution in kJ are shown in brackets

	Chloride	Hydroxide	Sulphate	Carbonate
Lithium	19.5 (−37)	5.3 (−21)	3.2 (−30)	0.2 (−18)
Sodium	6.2 (+4)	27.3 (−43)	1.4 (−2)	2.0 (−25)
Potassium	4.7 (+17)	20.0 (−55)	0.6 (+24)	8.1 (−33)
Rubidium	7.6 (+17)	17.4 (−63)	1.8 (+24)	19.6 (−40)
Caesium	11.0 (+18)	22.2 (−71)	4.9 (+17)	high (−53)
Beryllium	high	low	3.7	(unstable)
Magnesium	5.7 (−155)	low (+6)	2.8 (−91)	low (−25)
Calcium	6.7 (−83)	low (−16)	low (−18)	low (−12)
Strontium	3.4 (−52)	0.1 (−46)	low (−9)	low (−3)
Barium	1.7 (−13)	0.2 (−52)	low (+19)	low (+4)

		$\Delta H/\text{kJ mol}^{-1}$
vaporize lattice:	$Na^+OH^-(s) \rightarrow Na^+(g) + OH^-(g)$	+823
hydrate ions:	$Na^+(g) \rightarrow Na^+(aq)$	−406
	$OH^-(g) \rightarrow OH^-(aq)$	−460
overall:	$Na^+OH^-(s) \rightarrow Na^+(aq) + OH^-(aq)$	−43

Fig. 4.15 The dissolution of sodium hydroxide.

		$\Delta H/\text{kJ mol}^{-1}$
vaporize lattice:	$Mg^{2+}(OH)_2^-(s) \rightarrow Mg^+(g) + 2OH^-(g)$	+2846
hydrate ions:	$Mg^{2+}(g) \rightarrow Mg^{2+}(aq)$	−1920
	$2OH^-(g) \rightarrow 2OH^-(aq)$	−920
overall:	$Mg^{2-}(OH)^-(s) \rightarrow Mg^{2+}(aq) + 2OH^-(aq)$	+6

Fig. 4.16 The dissolution of magnesium hydroxide.

Lithium and beryllium chlorides are unusual in the *s*-block, being appreciably soluble in polar organic solvents such as alcohols. This behaviour indicates considerable covalent character in the bonding. In all ionic compounds there is a tendency for the cation to attract electrons from the anion. If the effect is large enough, this leads to partial sharing of electrons from the anion by the cation, that is, covalent character. The small sizes of Li^+ and Be^{2+}, and consequent high charge density, make them powerful electron attractors capable of deforming relative large anions such as Cl^-, Br^-, and I^- (Fig. 4.17).

The carbonates and nitrates of the *s*-block elements usually decompose on heating. In general, the alkali metal salts are more stable than those of the alkaline earths and thermal stability increases on passing down each group. For example, lithium carbonate and the Group 2 carbonates form the oxide and carbon dioxide on heating whilst the other Group 1 carbonates are thermally stable (Eqns 4.17 and 4.18).

electron drift

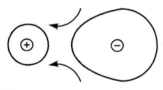

Fig. 4.17 Deformation of an anion by a cation.

$$Li_2CO_3(s) \rightarrow Li_2O(s) + CO_2(g) \tag{4.17}$$

$$MgCO_3(s) \rightarrow MgO(s) + CO_2(g) \tag{4.18}$$

In effect, such reactions simply involve the breakdown of the carbonate ion. This is promoted by the polarizing effect of the cation, which attracts electron density from the anion. If the cation is small enough, its charge density will cause sufficient polarization to break the bonds within the carbonate ion.

A similar effect occurs with the *s*-block nitrates. Only the nitrates of lithium and the alkaline earth metals decompose fully on heating to give the metal oxide (Eqns 4.19 and 4.20).

$$4LiNO_3(s) \rightarrow 2Li_2O(s) + 4NO_2(g) + O_2(g) \tag{4.19}$$

$$2Ca(NO_3)_2(s) \rightarrow 2CaO(s) + 4NO_2(g) + O_2(g)$$

(4.20)

The other alkali metal nitrates do decompose on heating but only as far as the *nitrite* and oxygen (Eqn 4.21).

$$2NaNO_3(s) \rightarrow 2NaNO_2(s) + O_2(g)$$

(4.21)

4.8 Lithium chloride: water absorption and rust protection

The cost of protecting the internal steel box-girder decks that make up the northern span of the spectacular Humber bridge in the UK by painting is estimated at about £5 million ($7.5 million). However, the Humber Bridge Board intends to stop corrosion using a desiccant dehumidifier based upon lithium chloride costing only £10 000 ($15 000). This system is already used to control the atmosphere inside the concrete anchor chambers which house the suspension cables at either end of the bridge.

Damp air inside the chambers is drawn through a slowly rotating wheel impregnated with lithium chloride. As the most soluble of the alkali metal chlorides, this compound readily absorbs moisture from the air. The moisture is removed by blowing warm air through the wheel and then expelled through a pipe to the outside of the bridge. The relative humidity inside the chambers is reduced to about 35%, which is low enough to stop the cables corroding.

5   *p*-Block elements

Table 5.1 Metals, non-metals (bold text), and metalloids (shaded boxes) in the *p*-block. The state of the non-solid elements at 25°C is also given

B	C	N	O	F	Ne
		gas	**gas**	**gas**	**gas**
Al	Si	P	S	Cl	Ar
				gas	**gas**
Ga	Ge	As	Se	Br	Kr
				liq.	**gas**
In	Sn	Sb	Te	I	Xe
					gas
Tl	Pb	Bi	Po	At	Rn
					gas

Elements of the *p*-block (Table 5.1) are those whose atoms have their last electron placed into a *p* subshell. They start with the Group 13 elements and conclude with the noble gas (Group 18) elements at which point the final electron completes the *p* level. Later sections in this chapter address trends in properties on moving down groups in the periodic table but it is appropriate here to consider trends on moving across the periodic table. For purposes of comparison of these horizontal trends, the *s*-block elements are included in the charts. The chemistry of the elements is dominated by electronic configuration. Each group in the periodic table has a group electronic configuration (Table 5.2).

Table 5.2 Group electronic configurations for the *s*- and *p*-block elements.

Group	1	2	13	14	15	16	17	18
Configuration	ns^1	ns^2	ns^2np^1	ns^2np^2	ns^2np^3	ns^2np^4	ns^2np^5	ns^2np^6

5.1   Trends across the periodic table

Some general periodic trends are outlined in Chapter 1. The *p*-block elements to the bottom left are metallic (Section 2.1) while those to the top right are non-metallic. A diagonal set of elements possess intermediate characteristics and are termed semi-metallic or metalloids (Table 5.1). Elements towards the top and right of the block are gases. Only bromine is a liquid at 25°C. The remainder are solids.

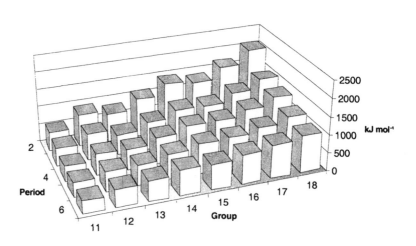

Fig. 5.1 Variation of first ionization energy for *s*- and *p*-block elements.

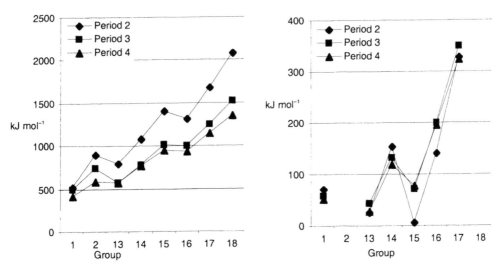

Fig. 5.2 Variation of first ionization energy (left) and electron affinity (right) for *s*- and *p*-block elements.

The ionization energy is the energy change associated with formation of ions M^+ from gaseous atoms M. Large positive numbers mean it is difficult to remove the electron. As shown by the charted first ionization energy (Fig. 5.1), it becomes progressively more difficult to remove an electron from the *s*- and *p*-block elements on moving from left to right across the table. In this direction, the nuclear charge increases and while the valence electrons are screened from the full charge by whatever core electrons there are, the resulting *effective nuclear charge* also increases. When the effective positive charge is larger, the valence electrons are held more tightly and more closely to the nucleus. This makes it more difficult to remove electrons from the atom. The plot in Fig. 5.1 also shows a trend for ionization energy to *decrease* on moving *down* the group. The valence orbitals are larger for the heavier elements and so electrons located within them are further away from the nucleus. They are also screened from the nucleus by the inner electrons and this means they are held less tightly and so are easier to remove.

The three-dimensional plot is very effective for seeing gross trends but a conventional plot reveals 'irregularities' more clearly (Fig. 5.2). This plot shows that the trend is not a smooth curve. Values for the Group 13 elements are lower than might have been expected, as are the values for the Group 16 elements. The group electronic configuration of the Group 13 elements is ns^2np^1. The outermost electron is in a *p* orbital and on average is further away from the nucleus than the *s* electrons. The *s* electrons also screen the *p* electron from the nucleus. The effect is that it is relatively a little easier to remove the *p* electron from an atom with an ns^2np^1 configuration than from an atom with the ns^2 configuration. A different argument is used to account for the dip at the Group 16 elements. Here the group electronic configuration is ns^2np^4. In these cases the final electron is located in an orbital that *already contains* an electron. This results in electron–electron repulsion in that *p* orbital. The effect of this repulsion is easier removal of that last electron.

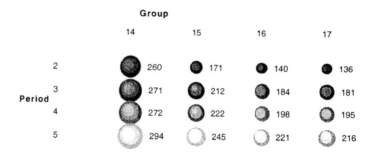

Fig. 5.3 Pauling ionic radii of the anions X⁻ (Group 17), X²⁻ (Group 16), X³⁻ (Group 15), and X⁴⁻ (Group 14).

For similar reasons, the tendency for the atoms to attract electrons to form anions increases across the table (Fig. 5.2). This means that the chemistry of elements to the right is dominated by tendencies to gain an electron and acquire a negative oxidation state rather than to lose an electron. This tendency declines towards the left and down the groups where positive oxidation states are more important. The electron affinity of the Group 15 element is very low — meaning that compared to the Groups 14 and 16 to either side, it is more difficult to form M⁻ ions. The group electronic configuration for Group 15 is ns^2np^3. Gain of an electron to form M⁻ gives ions with the configuration ns^2np^4. The fourth p electron is required to enter into a p-orbital that already contains an electron. There is an energy penalty to pay, however, since the two electrons mutually repel. This results in a much lower electron affinity.

5.2   Ionic size

In many instances, elements from the left of the periodic table form ionic compounds with anions of the Group 16 and 17 elements. The Pauling ionic radii are illustrated in Fig. 5.3 and show that the size of the Group 16 X²⁻ ion is about the same size as the adjacent Group 17 X⁻ ion. The ions get larger on descending the group because of orbital size and inner electron screening (see Section 5.1). These trends in size have ramifications for the structures of, for instance, the Group 1 halides, where the relative sizes of the anions and cations determine the solid state structure.

5.3   Trends in formulae of halides and oxides

The formulae of the Period 2 (Li–Ne) fluorides and the oxidation state of the s- and p-block elements are displayed in Table 5.3. Those of the Period 3 and 6 elements are shown in Tables 5.4 and 5.5. For Period 2 fluorides, the Group 14–18 elements are covalent and follow the octet rule. The Group 13 fluoride, BF_3, is covalent but boron has only six electrons in its outer shell (three from boron and one from each of the three fluorides). The fluorides of Group 1 and 2 elements are ionic lattices (indicated by the high melting

Table 5.3. The Period 2 fluorides

Oxidation number	1	2	13	14	15	16	17
+4				CF_4			
+3			BF_3		NF_3		
+2		BeF_2				OF_2	
+1	LiF						
0							F_2

Table 5.4. The Period 3 fluorides

Oxidation number	1	2	13	14	15	16	17
+6						SF_6	
+5					PF_5		ClF_5
+4				SiF_4		SF_4	
+3			AlF_3		PF_3		ClF_3
+2		MgF_2				$[SF_2]$	
+1	NaF						ClF
0							

Table 5.5. The Period 6 fluorides

Oxidation number	1	2	13	14	15	16	17
+6							
+5					BiF_5		
+4				PbF_4			
+3			TlF_3		BiF_3		
+2		BaF_2		PbF_2			
+1	CsF		TlF				
0							

Table 5.6. Melting points of the Period 2 fluorides

Formula	Melting point/°C
LiF	1118
BeF_2	1073
BF_3	144.5
CF_4	89
NF_3	66.5
OF_2	49.3
F_2	53.53

points in Table 5.6) with each ion possessing a stable noble gas electron structure — the fluoride always possesses the electron configuration of the *next* noble gas (Ne) while the cation always displays that of the *previous* noble gas (He). A similar pattern is observed for the Period 3 elements but note that SF_2 possesses only a fleeting existence.

To the right of the tables, 'octet expansion' is evident for the first time with PF_5, SF_4, and ClF_3 possessing ten electrons (because of the availability of *d* orbitals) in the valence shell of the central atom. The higher halides SF_6 and ClF_5 go further with 12 valence electrons. On descending the groups, structures such as TlF and PbF_2 are evident. These possess oxidation numbers for the *p*-block two less than the normal. It seems that the *s* electrons are less inclined to take part in bonding upon descending the groups. This is called the *inert pair effect*.

Recent calculations on element 114 (Uuq), which lies below lead in the periodic table, suggest that it might, in principle, be possible to make $UuqF_2$ but that $UuqF_4$ is much less likely.

5.4 Trends in properties of hydrides

The graph of boiling points of the *p*-block element hydrides (Fig. 5.4) shows that the boiling points of the period 2 hydrides of the Groups 15, 16, and 17 are higher than might have been expected. This is just as well, as it means

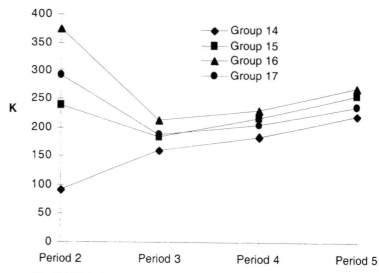

Fig. 5.4 Variation of boiling point of some main group species MH$_n$.

that water is normally liquid rather than a gas, allowing life as we know it to occur. These apparently anomalous boiling points are a consequence of hydrogen bonding. In water, electronegativity differences between the two elements mean that the O—H bonds are polarized so that the hydrogen possesses a slight positive charge and the oxygen is somewhat negatively charged. These partial charges are such that there are significant interactions between the partial charges upon adjacent molecules. Similar arguments may be used to explain the high values for the boiling points of NH$_3$ and HF.

5.5   Shapes of *p*-block molecules

VSEPR: Valence Shell Electron Pair Repulsion.

Clearly the shape of a molecule is important. It is important to know the shape of a molecule if one is to understand its reactions. It is also desirable to have a *simple* method to *predict* the shapes of molecules. For *p*-block compounds, the VSEPR method uses a simple set of electron accounting rules to carry out the prediction. Organic molecules are treated just as successfully as inorganic molecules.

Application of the VSEPR method requires some simplifying assumptions about the nature of the bonding. Despite these simplifications, the correct shape is nearly always obtained. In the VSEPR method as outlined here it is *assumed* that the geometry of a molecule depends *only* upon electron–electron interactions.

Some information is needed at the outset. The connectivity of the atoms in the molecule must be known, that is, which atoms are connected to which. The underlying assumptions made by the VSEPR method are the following:

- Atoms in a molecule are bound by electron pairs called *bonding pairs*. More than one set of bonding pairs of electrons may bind any two atoms together (multiple bonding).
- Some atoms in a molecule may also possess pairs of electrons not involved in bonding. These are called *lone pairs* or *non-bonded pairs*.
- The bonding pairs and lone pairs around any particular atom in a molecule adopt positions in which their mutual repulsions are minimized. The logic here is simple. Electron pairs are negatively charged and will get *as far apart from each other as possible*.
- Lone pairs occupy a larger solid angle at the nucleus than bonding electron pairs.
- Double bonds occupy a larger solid angle at the nucleus than single bonds.

Table 5.7 VSEPR geometries

Electron pairs	Geometry
2	linear
3	trigonal planar
4	tetrahedral
5	trigonal bipyramidal
6	octahedral

It is necessary to know the most favourable arrangement for any given number of electron sets surrounding any particular atom. These arrangements are found using simple geometrical constructions. This involves placing the nucleus of the atom in question at the centre of a sphere and then placing the electron pairs on the surface of the sphere so that they *are as far apart as possible*. The resulting arrangements are often intuitively obvious (Table 5.7).

For the case of just two electron pairs (Fig. 5.5) the arrangement is simple and the minimum energy configuration is when the electron pairs form a linear arrangement with the nucleus. In this configuration the electron pair–nucleus–electron pair angle is 180°. The coordination geometry of the central atom is described as *linear*.

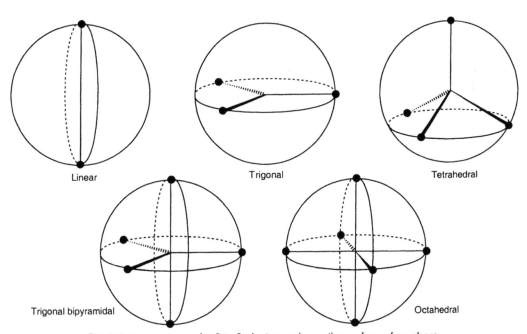

Linear   Trigonal   Tetrahedral

Trigonal bipyramidal   Octahedral

Fig. 5.5 Arrangements for 2 to 6-electron pairs on the surface of a sphere.

Three electron pairs arrange themselves *trigonally*, that is, with bond angles of 120°. For four electron pairs, one might expect the square-planar geometry to be favourable. However, tetrahedral bond angles are 109.5°, larger than the square-planar angles of 90°. The electron pairs are further apart in the tetrahedral arrangement than in a square–planar arrangement. There is more electron pair–electron pair repulsion in the square-planar geometry and so the *tetrahedral* geometry is favoured.

Five coordination is a little trickier. Most molecules whose shape is determined by five electron pairs are *trigonal bipyramidal*. There is another very reasonable candidate structure, and that is the square-based pyramid. In effect, this arrangement is an octahedron in which one group is removed and in which the four adjacent groups move down slightly to occupy partially the resulting vacancy. In practice, this geometry is only a little disfavoured relative to the trigonal bipyramid.

For six-coordinate systems, the *octahedral* geometry is by far the most important. An alternative geometry, the trigonal prism, is uncommon.

A Lewis single bond is always regarded as a σ bond. Recall that Lewis dot structures allow double and triple bonds between some elements. For VSEPR calculation purposes, a double bond is always regarded as a $\sigma + \pi$ bond interaction, while a triple bond is always treated as a $\sigma + 2\pi$ bond interaction. Double and triple bonds are treated as single sets of electrons and still *only occupy one coordination vertex*. The shape of a molecule is therefore dictated by the σ bond framework. Each vertex of the coordination polyhedron is necessarily occupied by a σ bond (possibly supported by π bonds) or a lone pair of σ symmetry, otherwise the vertex would not exist. The determination of a molecule's geometry therefore reduces to a calculation of the number of electrons contained in the σ orbitals. Electrons in π bonds must therefore be recognized and care taken not to confuse them with σ bond electrons. The following procedure, which is not unique, works well. The examples which follow will make the procedure clear.

There are two environments in a trigonal bipyramid, axial and equatorial. These two environments are chemically distinct.

1. Draw a Lewis structure for the molecule or ion involved. In the following examples it will become clear that some simplifications are possible.
2. Determine the number of valence electrons on the neutral central atom.
3. Write down a modified Lewis structure by assigning all atoms or groups bonded to the atom in question as *singly, doubly,* or *triply* bonded. Example assignments are given in Table 5.8. It is not necessary to write in lone pairs, the calculation will determine these. Assign all singly bonded groups as shared electron pair bond types, with the exceptions of dative bound groups discussed below. Always regard groups such as =O and =S as double bonded to the central atom with the double bond consisting of a σ and a π bond, both of which are electron pair bonds. Regard groups such as $\equiv$N and $\equiv$P as always triple bonded, with the triple bond consisting of a σ and two π bonds, all of which are shared electron pair bonds.

Table 5.8. Classification of formal bonding type to central atom

Single	Double	Triple
F, Cl, Br, I		
OH, SH	=O, =S	
NH_2	=NH, =PH	$\equiv$N, $\equiv$P
Me, Ph	$=CH_2$	$\equiv$CH
H		
$SiMe_3$		

4. The coordination geometry is dictated by the σ framework only. It is now necessary to discount those central atom electrons that are involved in π bonds. Since each π bond is a shared electron pair with one electron arising from each atom, subtract one electron for each π bond involving the central atom.

5. Any overall charge on the molecule is *always* assigned to the central atom, even if later reflection requires that it may be best to assign it elsewhere. Thus, a negative charge constitutes an additional electron for the central atom, while a positive charge requires subtraction of one electron from the central atom electron count.

6. Divide the total number of electrons associated with the σ framework by 2 to give the number of σ electron pairs. Assign a coordination geometry and perhaps distinguish between isomers.

A statement of rules always seems more imposing than their actual operation. It is therefore appropriate to examine some sample calculations. The calculation for methane shows that the carbon atom is associated with 8 electrons in the σ framework. This corresponds to four shape-determining electron pairs. The coordination geometry of carbon is consequently tetrahedral. There are four bonded groups, therefore there are no lone pairs.

Ammonia also has four electron pairs and the *coordination geometry* of nitrogen is based upon a tetrahedral arrangement of electron pairs. There are

---

Methane, CH₄

Lewis structure:
```
      H
      |
  H — C — H
      |
      H
```

central atom: **carbon**

valence electrons on central atom:	4
4 H each contribute 1 electron:	4
total:	8
divide by 2 to give electron pairs:	4

4 electron pairs: *tetrahedral* geometry

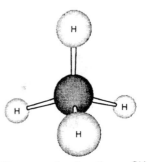

The geometry of methane, CH₄

Ammonia, NH₃

Lewis structure:
```
  H — N — H
      |
      H
```

central atom: **nitrogen**

valence electrons on central atom:	5
3 H each contribute 1 electron:	3
total:	8
divide by 2 to give electron pairs:	4

4 electron pairs: *tetrahedral* arrangement

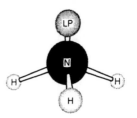

The geometry of ammonia, NH₃

three bonded groups, therefore there is one lone pair. Although lone pairs help determine shape, they are not described normally as part of the shape. So, the *shape* of ammonia is *pyramidal*.

In a bonding pair of electrons, the two electrons are located between two nuclei, and are attracted by both. A lone pair is different. It is necessarily only attracted to one nucleus and the consequence is that it adopts a position effectively *closer* to that one nucleus than the bonding pairs of electrons. The effect is for the lone pair to repel other electrons around the central atom more strongly. This means that the effective solid angle occupied by a lone pair is *greater* than that occupied by a bond pair. Lone pairs demand greater angular room, and are located closer to their atoms than bond pairs. The consequence of this for ammonia is that the lone pair makes room for itself by pushing the three hydrogen atoms together a little and the H—N—H bond angles are slightly less (106.6°) than the ideal tetrahedral angle of 109.5°.

Water has four electron pairs and the *coordination geometry* of oxygen is based upon a tetrahedral arrangement of electron pairs. Since there are only two bonded groups, there are two lone pairs. Since the lone pairs are not included in the description, the *shape* of water is bent. The lone pairs close the H—O—H bond angle from the ideal tetrahedral angle to 104.5°.

Boron trifluoride only has six valence electrons and is one of the relatively rare second period covalent molecules that disobeys the octet rule. There are three bonded groups and so no lone pairs. Six electrons implies three electron pairs and therefore a *trigonal* geometry.

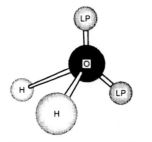

The geometry of water, OH_2

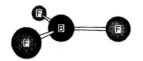

The geometry of boron trifluoride, BF_3

Water, OH_2

Lewis structure: H—O—H

central atom: **oxygen**	
valence electrons on central atom:	6
2 H each contribute 1 electron:	2
total:	8
divide by 2 to give electron pairs:	4
4 electron pairs: *tetrahedral* arrangement	

Boron trifluoride, BF_3

Lewis structure:
$$F-B-F$$
$$|$$
$$F$$

central atom: **boron**	
valence electrons on central atom:	3
3 F each contribute 1 electron:	3
total:	6
divide by 2 to give electron pairs:	3
3 electron pairs: *trigonal* geometry	

Hexafluorophosphate, [PF₆]⁻

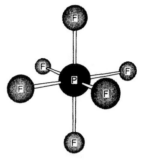

Lewis structure:

central atom: **phosphorus**

valence electrons on central atom	5
6 F each contribute 1 electron	6
add one for the negative charge on P	1
total:	12
divide by 2 to give electron pairs:	6
6 electron pairs: octahedral geometry	

The geometry of hexafluorophosphate, [PF₆]⁻

Since BF_3 is two electrons short of the octet configuration, a characteristic reaction is to react with lone pair donors. Acceptance of a pair of electrons makes BF_3 a *Lewis acid*. The electron pair of, say, NH_3, is a *Lewis base*.

For hexafluorophosphate, [PF₆]⁻, there are six bonded groups and so no lone pairs. This anion is useful in synthesis since it often aids the crystallization of bulky cations by providing a reasonable size match for the cation. Note that the negative charge for the purposes of the calculation is placed on phosphorus for the purpose of the calculation even though the negative charge is in reality delocalized over all seven atoms of the ion.

There are four σ-bonded hydrogen groups attached to nitrogen in the ammonium cation and the calculation predicts no lone pairs on the central atom. Note that the positive charge is placed formally on nitrogen, even though it is in reality delocalized over the whole ion.

For sulphur dioxide, SO_2, the calculation predicts one lone pair on the central sulphur. Here the VSEPR treatment is seen for a double bond. In this case the competition between the lone pair and the two double bonds is a draw in the sense that the O=S=O bond angle is very close to the ideal angle of 120° at 119.3°. It is interesting to compare this to the structure of nitrogen

Ammonium, [NH₄]⁺

Lewis structure:

central atom: **nitrogen**

valence electrons on central atom:	5
4 H each contribute 1 electron:	4
subtract one for the positive charge on N:	−1
total:	8
divide by 2 to give electron pairs:	4
4 electron pairs: *tetrahedral* geometry	

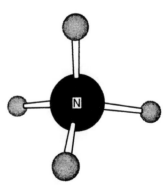

The geometry of the ammonium cation, [NH₄]⁺.

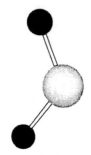

The geometry of SO₂

Sulphur dioxide , SO₂

Lewis structure: O=S=O

central atom: **sulphur:**

valence electrons on central atom:	6
2 O each contribute 1 electron:	2
subtract one for each electron contributed by S to the π bond:	−2
total:	6
divide by 2 to give electron pairs:	3

3 electron pairs: trigonal geometry

---

dioxide, NO_2. The shape of nitrogen dioxide is determined by one less electron than that of SO_2 and instead of a lone pair there is a lone electron (in effect, half a lone pair). As a consequence, that half-filled orbital repels less than a filled lone pair orbital and the ONO angle opens out to 134°. Addition of an electron to NO_2 to make NO_2^- results in completion of the lone pair and so the ONO angle closes down, in this case to less than 120° (actually 115°).

5.6   Group 13 elements

Group properties

Strongly electronegative and reactive diatomic elements

Group name: none

Group members: B, Al, Ga, In, Tl

Group configuration ns^2np^1

The group 13 elements do not carry any special group name. All the elements are important for various reasons. See Fig. 5.6 for group trends. This group is the most metallic in nature of the *p*-block elements and only boron is non-metallic.

Boron exists as a number of complex allotropic forms. For instance, α- and β-rhombohedral boron contain B_{12} icosahedra linked through B—B interactions. The most common mineral source of boron is the aluminosilicate tourmaline. Aluminium is a structurally useful metal that would normally be highly reactive. However, aluminium surfaces are normally protected by an inert oxide layer. Coloured aluminium surfaces are possible through the addition of pigments within the oxide layer. Aluminium is mined in huge scales as bauxite ($Al_2O_3 \cdot 2H_2O$). Bauxite contains Fe_2O_3 and SiO_2 impurities. In order to provide pure aluminium, these impurities must be

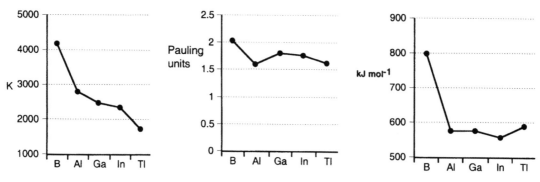

Fig. 5.6 Some group trends of the Group 13 elements: boiling point (left), Pauling electronegativity (centre), and the first ionization energy (right).

removed from the bauxite. This is done by treatment with sodium hydroxide solution, which results in a solution of sodium aluminate and sodium silicate. The iron remains behind as a solid. When CO_2 is blown through the resulting solution, the sodium silicate stays in solution while the aluminium is precipitated out as aluminium hydroxide. The hydroxide can be filtered off, washed, and heated to form pure alumina, Al_2O_3. Pure aluminium is obtained from the pure Al_2O_3 by an electrolytic method, necessary as aluminium is rather electropositive.

Gallium is notable for its melting point, around 30°C, or barely above room temperature. Gallium will melt when warmed by the hand. The boiling point is, however, high at about 2200°C giving gallium the greatest liquid range of any element.

All the elements other than boron form M^{3+} ions and all form covalent bonds in various MX_3 molecules. The electron count in molecules such as BF_3 and $AlCl_3$ is only six, meaning the octet structure is not achieved. The chemistries of such species are dominated by their tendencies to acquire an extra pair of electrons to achieve that octet structure. In BF_3, the boron is trigonal (Fig. 5.7), as predicted by VSEPR methods. Each of the three B—F σ bonds contains two electrons, six in total. Three of these originate from boron, and one each from the three fluorine atoms. Boron has a *vacant p* orbital lying out of the plane of the molecule. This orbital interacts with the electron density from the fluorine lone pairs and in effect the fluorine atoms form slight dative interactions into the vacant *p* orbital. This means that the effective bond order between boron and each fluorine atom is a little greater than 1, accounting for the rather short B—F bonds actually observed.

The vapour phase structure of $AlCl_3$ is interesting. The electron count for aluminium in $AlCl_3$ is six, two short of an octet. This makes aluminium in $AlCl_3$ inclined to accept an electron pair, in other words to behave as a Lewis acid. There are plenty of electron pairs on the chlorine atoms and aluminium achieves the octet count by accepting an electron pair from a neighbouring $AlCl_3$. In turn, the first $AlCl_3$ group donates an electron pair to the second aluminium atom. The resulting dimer (Fig. 5.8) has two chlorine bridges.

Alumina, Al_2O_3, is most important, largely since it is the main source for aluminium metal. In corundum, the Al_2O_3 lattice (Fig. 5.9) may be regarded

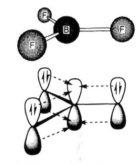

Fig. 5.7 Boron trifluoride, BF_3, and orbital interactions that lead to a shortening of the B—F bond.

Lewis acid: electron acceptor
Lewis base: electron donor

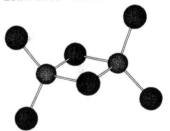

Fig. 5.8 The molecular structure of $AlCl_3$.

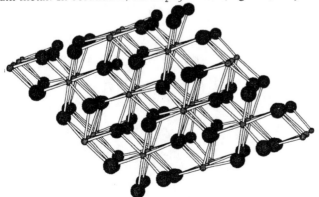

Fig 5.9 The solid state structure of alumina, Al_2O_3.

as ionic, that is, consisting of Al^{3+} and O^{2-} ions. In support of this notion, molten alumina conducts electricity. However, the oxide ions are distorted by the small, highly charged aluminium tri-cations resulting in a flow of electron density back towards the cations. In effect, this is a covalent contribution. Very crystalline samples of corundum are clear and if there are traces of *d*-block cations such as iron or titanium present as impurities, the result is the gemstone blue sapphire. If the impurity is chromium, the resulting gemstone is red and is called ruby (Chapter 6).

Lower down the group there is a tendency for elements such as thallium to display oxidation number +1 as well as +3. So, $TlCl_3$ loses chlorine upon warming to form $TlCl$. The Tl(I) oxidation state is more stable than Tl(III).

5.7   Group 14 Elements

The Group 14 elements do not carry any special group name. All the elements are important for various reasons.

The most common form of pure carbon is graphite. Diamond is a second form (allotrope) of carbon but is much less common. Graphite possesses a layer structure (Fig. 5.10) in which each carbon is directly bound to three other carbon atoms. Delocalization in the bonding is evident since the C—C distances are equal and shorter than normal single bonds. There are two forms of graphite related by the way the layers stack upon each other. 'Amorphous' forms of carbon such as soot and lampblack are materials consisting of very small particles of graphite.

Diamond is a slightly more compact structure, hence its greater density. The appearance of diamond is well known and it is also one of the hardest materials known. Like graphite, it is relatively unreactive but does burn in air at 600–800°C. Each carbon atom is bound to four neighbours in a tetrahedral fashion and so each diamond crystal is a single giant lattice structure. In principle (and in practice!) graphite may be converted into diamond by the application of heat and pressure.

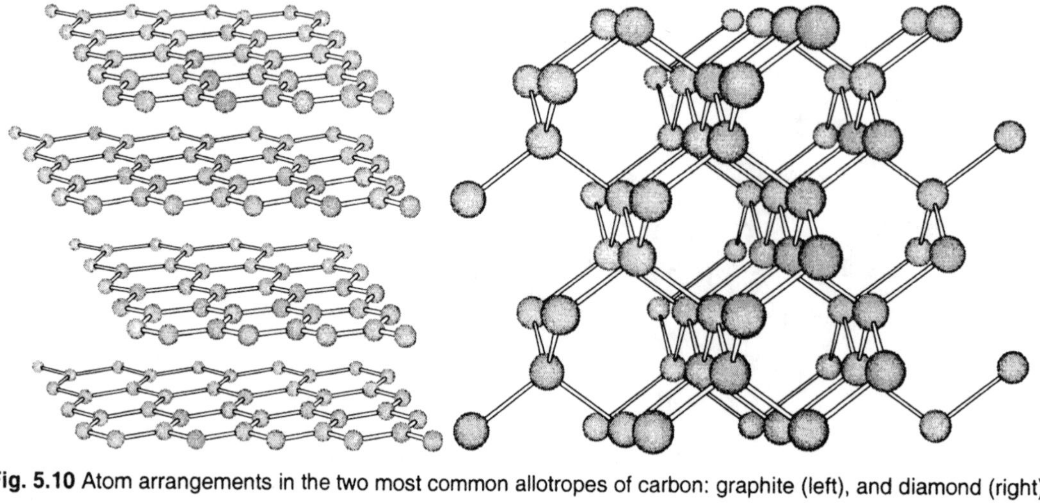

Group properties

Strongly electronegative and reactive diatomic elements

Group name: none

Group members: C, Si, Ge, Sn, Pb

Group configuration ns^2np^2

Fig. 5.10 Atom arrangements in the two most common allotropes of carbon: graphite (left), and diamond (right).

Recently another allotrope of carbon was characterized. Whereas diamond and graphite are infinite lattices, buckminsterfullerene, C_{60}, is a discrete molecular species (Fig. 5.11). The buckminsterfullerene molecule is a net of pentagons and hexagons folded into a sphere. The effect is very similar to the patchwork of pentagonal and hexagonal pieces of leather that sewn together make up an association football (soccer ball). Buckminsterfullerene is now commercially available and has also been identified in interstellar space and soot. Other fullerenes are now known as well.

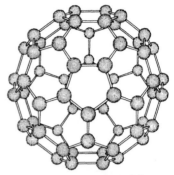

Fig. 5.11 C_{60}, buckminsterfullerene.

Forms of the heavier elements corresponding to graphite are not known and the structures of silicon, germanium, and grey tin are related to the diamond structure. Reflecting the normal tendency for metallic character to increase down the group, the structure of lead is cubic close packed and is metallic.

Whereas carbon dioxide, CO_2, is a gas, the heavier analogue, SiO_2, is known as quartz or crystobalite, or common sand. Carbon dioxide is a linear triatomic molecule. Silica's solid state structure is shown in Fig. 5.12 and shows that each silicon atom is surrounded tetrahedrally by four oxygen atoms. Overall the structure is a giant covalent lattice (or macromolecule).

Silicon oxide is important for the manufacture of glass. Soda glass is made by melting silica (SiO_2, sand), together with sodium carbonate, Na_2CO_3, and limstone, $CaCO_3$, at about 1800 K (Eqns 5.1 and 5.2). On cooling, a transparent material made up from calcium and sodium silicates results and is called soda glass.

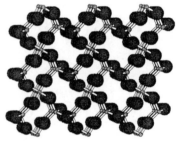

Fig. 5.12 Silica, SiO_2.

$$Na_2CO_3 + SiO_2 \rightarrow Na_2SiO_3 + CO_2 \qquad (5.1)$$

$$CaCO_3 + SiO_2 \rightarrow CaSiO_3 + CO_2 \qquad (5.2)$$

Today, enormous quantities of this type of glass are made every year and there are many other forms in use as well. For instance, addition of boron oxides to the melt gives borosilicate glasses (Pyrex) which can withstand high temperatures. Coloured glasses are made by the addition of metal oxides to the melt. Thus, addition of certain cobalt salts to the melt gives vivid blue

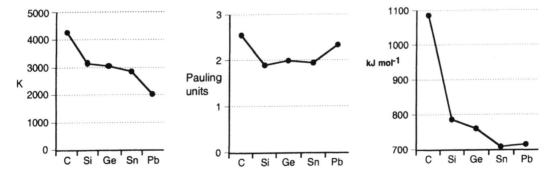

Fig. 5.13 Some group trends of the Group 14 elements: boiling point (left), Pauling electronegativity (centre), and the first ionization energy (right).

colours. Red glass is produced by the addition of cadmium sulphide and selenide. Most glass viewed edge-on is pale green. This is because of Fe(II) impurities. Glass with enhanced Fe(II) concentrations are very distinctly green, and such glasses possess very desirable thermal properties.

Group properties

Strongly electronegative and reactive diatomic elements
Group name: pnictogens
Group members: N, P, As, Sb, Bi
Group configuration ns^2np^3

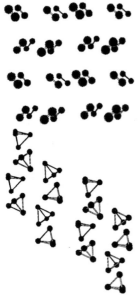

Fig. 5.14 The solid state structure of N_2 (top) and white phosphorus, below.

5.8 Group 15 elements

The Group 15 elements are sometimes referred to as *pnictogens*. All the elements are important and bismuth is notable for being the heaviest element with non-radioactive isotopes.

Nitrogen as N_2 (dinitrogen) makes up about four-fifths of the atmosphere and while not utilized directly during breathing is important since it dilutes the oxygen component of the atmosphere to acceptable levels. Nitrogen gas is diatomic with a formal triple bond connecting the two atoms. The Lewis representation shows that the triple bond is required to allow an octet of electrons to form at each atom and suggests correctly that N_2 should not contain any unpaired electrons — that is, it is diamagnetic. In the solid state the N_2 molecules (Fig. 5.14) stack with a characteristic herring-bone pattern.

Phosphorus is very different from nitrogen. It exists as a number of allotropes including red, white, and black. However, there are a number of further modifications and the nature of some of these is still unclear. In the vapour phase below 800°C, phosphorus exists largely as tetrahedral P_4 and these units are also present in solid white phosphorus (Fig. 5.14). The nature of red and black phosphorus seems much more complex but both forms are stable in air, whereas white phosphorus must be stored under water since it ignites readily.

The graphs in Fig. 5.15 indicate that nitrogen stands out from the other members of the group, all of which show chemistries quite closely related to each other. This phenomenon is also evident in some other *p* block groups. It is not unusual for textbooks to treat the chemistry of nitrogen in a different section to that of the lower elements. Nitrogen is usually trivalent whereas the lower elements display variable valencies. Trends towards more metallic behaviour increases down the group. Three electrons are required for these elements to acquire an octet structure. However, the high energy cost in

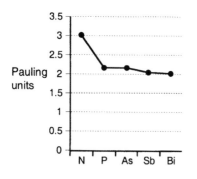
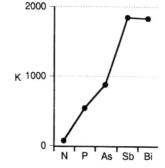
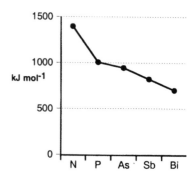

Fig. 5.15 Some group trends of the Group 15 elements: boiling point (left), Pauling electronegativity (centre), and the first ionization energy (right).

Fig. 5.16 NH_3 $\angle H—N—H = 106.7°$ (left); PH_3 $\angle H—P—H = 93.4°$ (centre left); AsH_3 $\angle H—As—H = 92.1°$ (centre right); SbH_3 $\angle H—Sb—H = 91.6°$ (right).

achieving the M^{3-} ion structure means that M^{3-} ions are rare, instead covalency is the norm. Closed electron shell configurations could also, in principle, be achieved through formation of M^{5+} but these are not known. Salts containing Bi^{3+}, however, are known.

The binary hydrogen compounds are all pyramidal. While VSEPR rules indicate all structures are based upon four electron pairs, one of which is a lone pair, only ammonia has bond angles corresponding to a tetrahedral arrangement of electron pairs (Fig. 5.16). The bond angles in the lower forms of MH_3 are close to 90°. One way to rationalize this is the suggestion that the lone pairs in the heavier compounds are located closer to the central atom, forcing the three hydrogen atoms closer together.

The Group 15 elements form a number of important oxides. Those of nitrogen include nitrous oxide (N_2O), nitric oxide (NO), nitrogen dioxide (NO_2), and dinitrogen tetroxide (N_2O_4). Nitrous oxide is a gas and is relatively unreactive. It is known for its anaesthetic properties ('laughing gas'). Nitric oxide is formed in many reactions, including those involving dissolution of metals in concentrated nitric acid.

$$8HNO_3 + 3Cu \rightarrow 3Cu(NO_3)_2 + 4H_2O + 2NO(g) \qquad (5.3)$$

Nitric oxide possesses an odd number of electrons [N (5) + O (6) = NO (11)]. It is therefore *paramagnetic*. The odd electron is removed quite easily and the resulting nitrosonium ion (NO^+) is well known, in particular for its ability to coordinate as a ligand to *d*-block elements. Nitric oxide reacts very easily with oxygen to form the well–known brown nitrogen dioxide, NO_2. Nitrogen dioxide also has an odd number of electrons and is also, therefore, paramagnetic. In this case, two molecules of NO_2 combine through pairing of the odd electron to form colouress N_2O_4.

Paramagnetic compounds are those with at least one unpaired electron. *Diamagnetic* compounds are those with no unpaired electrons.

$$2NO_2 \underset{\text{heat}}{\overset{\text{cool}}{\rightleftharpoons}} N_2O_4 \qquad (5.4)$$

This contains no unpaired electrons and is therefore diamagnetic. Nitrogen dioxide and N_2O_4 are present in temperature–dependent equilibrium. On cooling to the solid, the equilibrium lies completely towards N_2O_4. In the vapour phase above 140°C, the equilibrium lies completely towards NO_2, while in cooler vapours and in the liquid phase there are mixtures.

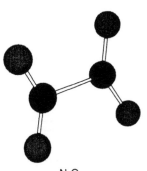

N_2O_4

LIVERPOOL JOHN MOORES UNIVERSITY
LEARNING SERVICES

The most important phosphorus oxide is P_4O_{10} (phosphorus(V) oxide, referred to as phosphorus pentoxide, P_2O_5, in older literature). This consists of discrete tetrahedral molecules (Fig. 5.17). Each edge of the tetrahedron consists of POP links and each P atom is also bonded to another oxygen atom. A related molecule P_4O_6, phosphorus(III) oxide, is also known but does not contain the terminal oxygen atoms. The compound P_4O_{10} is an excellent drying agent, even to the extent of removal of H_2O from sulphuric acid to form SO_3. Hydration of P_4O_{10} results in H_3PO_4, orthophosphoric acid, or just phosphoric acid. This is an important acid and is commonly available as a syrupy 85% pure form containing some water. Phosphoric acid is *tribasic*, and phosphates containing tetrahedral phosphate ions, PO_4^{3-}, are well known and important. Ammonium phosphate is an important fertilizer for example.

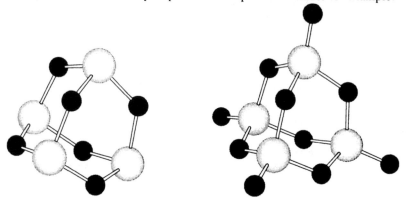

Fig. 5.17 The structures of P_4O_6 (left) and P_4O_{10} (right).

5.9   Group 16 elements

Group properties

Strongly electronegative and reactive elements

Group name: chalcogens

Group members: O, S, Se, Te, Po

Group configuration ns^2np^4

The Group 16 elements are often referred to as *chalcogens*. All are important except polonium, which is very rare. As for the Group 15 elements, the properties of the first element, oxygen, are somewhat distinct from those of the lower elements (see Fig. 5.18).

Oxygen as O_2 (dioxygen) makes up about a fifth of the atmosphere and is

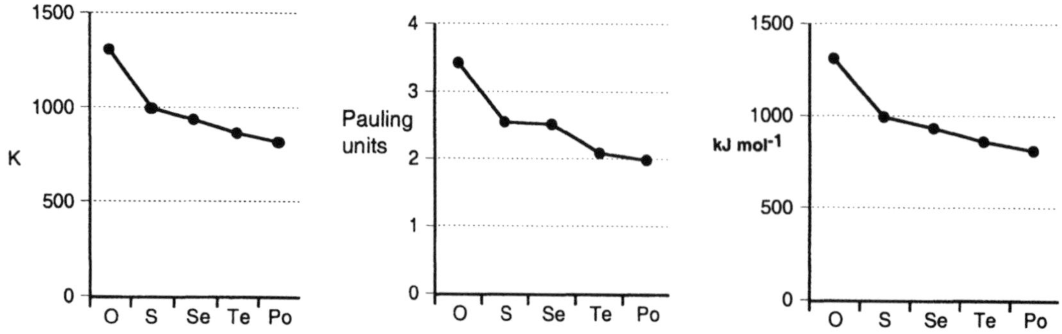

Fig. 5.18 Some group trends of the Group 16 elements: boiling point (left), Pauling electronegativity (centre), and the first ionization energy (right).

fundamental to life. Dissolved oxygen is also abundant in water. Most of this oxygen is biological in origin, specifically from the photosynthetic transformation of water into oxygen by green plants. Most oxygen for laboratory use is obtained through the liquefaction of air and is supplied in cylinders under pressure. On a small scale, oxygen can be generated by the thermal decomposition of potassium chlorate (Eqn 5.5).

$$2KClO_3 \xrightarrow{\Delta} 2KCl + 3O_2 \qquad (5.5)$$

Oxygen is diatomic with a formal double bond connecting the two atoms. The Lewis representation shows that the double bond is required to allow an octet of electrons to form at each atom but suggests that O_2 should not contain any unpaired electrons (compounds containing no unpaired electrons are referred to as diamagnetic). In fact, O_2 is *paramagnetic*, that is, it contains unpaired electrons. More sophisticated views of the bonding than the Lewis model successfully account for this phenomenon. Liquefied and solid oxygen are both pale blue.

A second form (*allotrope*) of oxygen is called *ozone* and has the formula O_3. Ozone is paramagnetic and is blue when in a liquid form. Pure ozone is explosive. The solid is violet–black. While ozone is toxic, it is important in the upper atmosphere as the ozone layer. The ozone layer has a role in the absorption of harmful ultraviolet wavelengths (220–290 nm) and there is currently great concern about emerging 'holes' in the ozone layer over the poles which seem to lead to increases in skin cancer rates. In part, the holes in the ozone layer are caused by release of refrigerants and aerosol propellants such as CF_2Cl_2 and $CFCl_3$ into the atmosphere, and by nitrogen oxides from engine exhausts. Once in the upper atmosphere these result in radicals such as $Cl^{\cdot}$ which can catalyse ozone decomposition (Eqns 5.6–5.7) – the $^{\cdot}O^{\cdot}$ radical is produced through photodissociation of dioxygen.

$$Cl^{\cdot} + O_3 \rightarrow O_2 + ClO^{\cdot} \qquad (5.6)$$

$$ClO^{\cdot} + {}^{\cdot}O^{\cdot} \rightarrow O_2 + Cl^{\cdot} \qquad (5.7)$$

Sulphur is found in nature as the free element and in ores. It is an important and undesirable impurity in coal and crude oils from which it should be removed. Elemental sulphur exists as a number of allotropes, in particular various forms of S_8, although larger rings such as S_{12}, S_{20}, etc. are known (Fig. 5.19). Selenium and tellurium are usually impurities in sulphur ores while there are no known stable isotopes of polonium. As always, the trend towards metallic properties increases down the group and polonium can safely be regarded as metallic. For example, the oxide PoO_2 is ionic and reacts with HCl to form $PoCl_4$.

Oxygen, as implied by its name, is a good oxidizing agent and oxides of nearly all the elements are known. Sulphur dioxide, SO_2, and sulphur trioxide, SO_3, are particularly important. The former is a pollutant produced through burning sulphur-containing fuels such as coal and some oils. Once in

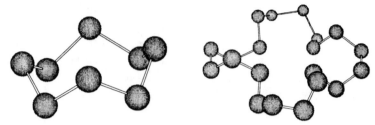

Fig 5.19 The solid state structures of S_8 and S_{20}.

the air it reacts with oxygen and water to form sulphuric acid, H_2SO_4, which dissolves in clouds, causing acid rain. Acid rain is a major political issue and while technology exists to reduce SO_2 emissions, its use is far from universal. Sulphur dioxide, SO_2, reacts with oxygen under the influence of a catalyst such as platinum or V_2O_5, to form SO_3. In industrial plants the SO_3 is absorbed into 98% sulphuric acid to form fuming sulphuric acid, oleum, which is then diluted back with water to form pure sulphuric acid. This is safer than the direct reaction of SO_3 with water since that reaction is too exothermic for convenience. Sulphuric acid is the chemical produced in greatest tonnage every year around the world.

Whereas the binary hydrogen oxide H_2O, water, is obligatory for life on this planet, the corresponding sulphur compound, H_2S, is not. Indeed, the foul smelling dihydrogen sulphide is more toxic than hydrogen cyanide, HCN. The H—O—H bond angle is 104.5°, that is, slightly closed from the ideal tetrahedral angle of 109.5° as a result of repulsive effects of the lone pairs. However, the corresponding angles for the lower analogues approach 90° (H—S—H, 92°, H—Se—H, 91°, and H—Te—H, 90°). The closeness of these values to 90° is not treated satisfactorily by simple VSEPR rules but can be rationalized in the same way as the corresponding values for the group 15 elements (Section 5.8). The thermal stability of the compounds decreases down the group as the bond energies decrease. The graph of boiling points for H_2O–H_2Te shows the expected increase on going down the group, with the clear exception that the boiling point of water is far higher than might have been expected. This is fortunate as it means that liquid water is available at temperatures convenient for life. The origin of this phenomenon is hydrogen bonding (Section 3.2).

The graphs in Fig. 5.18 show that oxygen is quite distinct from the lower elements, and that the properties of the lower elements are somewhat similar to each other. The decreasing electronegativity down the group implies that the lower elements are less likely to form ionic compounds. Higher oxidation states and higher coordination numbers are evident down the group. This 'octet expansion' after period 2 allows for the existence of species such as SF_6 and SF_4O.

Group properties

Strongly electronegative and reactive diatomic elements

Group name: halogens

Group members: F, Cl, Br, I, At

Group configuration ns^2np^1

5.10 Group 17 elements

The Group 17 elements are often referred to as *halogens*. They exist as diatomic molecules but none is found uncombined in nature. All are important except astatine, which is virtually absent from the planet. What little is known about its properties indicate that they extrapolate from those of the lighter halogens.

Fluorine is found in nature as minerals such as flurospar (Fig. 5.20, CaF_2). It may be recovered with difficulty as a highly reactive and corrosive pale yellow gas by electrolysis of hot molten mixtures of KF and HF. It is difficult to store as it reacts with most materials but steel and *Monel metal* containers may be used as the metal surfaces deactivate through the formation of unreactive surface fluorides. Volatile fluorine compounds such as CCl_2F_2 are used as refrigerants. Polymerized $CF_2{=}CF_2$ (polytetrafluoroethene, PTFE) is resistant to attack by most chemicals, insulates well, and provides a non-stick surface (such as in non-stick cooking pans). Fluorine reacts with water to form HF and oxygen (Eqn 5.8).

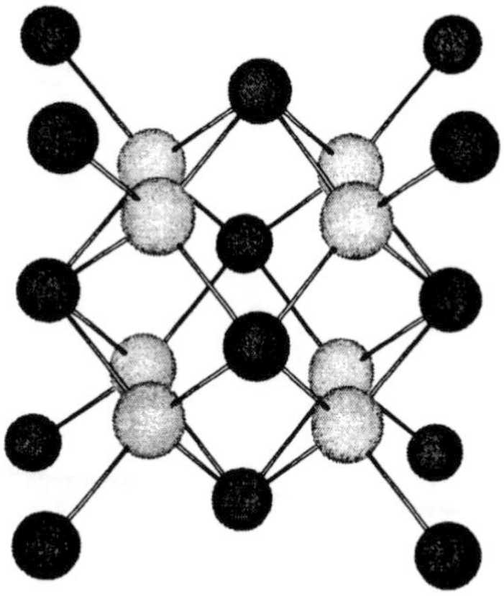

$$2F_2 \text{ (g)} + H_2O \text{ (l)} \rightarrow 4HF \text{ (aq)} + O_2 \text{ (g)} \tag{5.8}$$

Chlorine is found largely in seawater where it exists as chloride ions. It is recovered as reactive, corrosive, pale green chlorine gas from brine (a solution of sodium chloride in water) by electrolyis (Eqn 5.9). It is reasonably soluble in water but reacts further to form HOCl (Eqn 5.10).

$$Na^+ + Cl^- + H_2O \rightarrow Na^+ + {}^1\!/_2Cl_2 + {}^1\!/_2H_2 + OH^- \tag{5.9}$$

$$Cl_2 \text{ (g)} \rightarrow Cl_2 \text{ (aq)} \rightleftharpoons H^+ + Cl^- + HOCl \tag{5.10}$$

The reaction with base is faster and results in OCl^-; however, this tends to react further via a disproportionation reaction to form chlorate, ClO_3^- (Eqn 5.11). Chlorates such as sodium chlorate are weed killers. With care, because of the danger of explosions, chlorates may be converted into perchlorates, ClO_4^-. Perchlorates are sometimes used in fireworks but they are often percussion sensitive, requiring careful handling.

$$3Cl_2 \text{ (g)} + 6OH^- \text{ (aq)} \rightarrow 3Cl^- + 3OCl^- + 3H_2O \rightarrow 5Cl^- + ClO_3^- + 3H_2O \tag{5.11}$$

Fig. 5.20 The structure of CaF_2, fluorite.

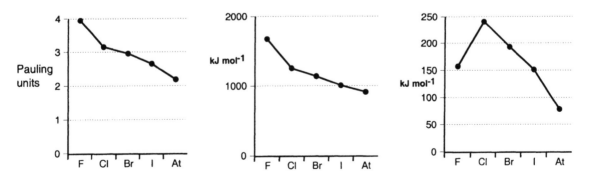

Fig. 5.21 Some group trends of the Group 17 elements: Pauling electronegativity (left), the first ionization energy (centre,), and bond enthalpies in the halogens X_2 (right).

Bromine also occurs in seawater as the sodium salt but in much smaller quantities than chloride. It is recoverable through the treatment of seawater with chlorine gas and flushing through with air. The principle of oxidation of bromide to bromine is shown by the addition of a little chlorine water to aqueous solutions of bromide. These become brown as elemental bromine forms. It is a corrosive, volatile dark-red mobile liquid at room temperature. Bromine is reasonably soluble in water and very soluble in a number of organic solvents.

Iodine also occurs in seawater and some marine life concentrates iodine. It also may be isolated by oxidation. It is a nearly black crystalline solid which must be stored in a sealed vessel as it sublimes at room temperature to give a characteristic corrosive violet vapour. While slightly soluble in water it is quite soluble in some organic solvents. Solutions in alcohol (*tincture of iodine*) are antiseptic and so useful for minor cut and grazes.

The halogens are all a single electron short of filled valence electron configurations and the tendency for them to acquire that electron dominates their chemistry. This is done either by acquisition of an electron to form the anion X^- or through formation of a single covalent bond. All the halogens are highly electronegative — fluorine the most so (Fig. 5.21). Fluorine itself is particularly reactive. This is because of the weakness of the F—F bond (Fig. 5.21) and the strength of M—F bonds. The weakness of the F—F bond is connected with lone pair–lone pair repulsions between the adjacent atoms in difluorine.

Most halide salts are soluble in water, those of Pb(II) and Ag(I) are exceptions. Thus halide ions in water may be tested for by the addition of a little aqueous silver nitrate. This gives white to yellow (AgCl is white, AgBr is cream, and AgI is yellow) precipitates except in the case of fluoride. All four halide ions precipitate with Pb(II).

Silver bromide is important in black and white photography and is a component of the film. Silver bromide decomposes to silver on exposure to light. Little light is required as the silver atoms produced induce the formation of many more silver atoms. Part of the development process involves treatment (*fixing*) with silver thiosulphate (*hypo*). This results in the removal of excess AgBr as the soluble ion $[Ag(S_2O_3)_2]^{3-}$. The remaining silver stains the film producing the negative.

Most elements in the periodic table form halides. Those of hydrogen are particularly important. Halides vary in nature from highly ionic (the Group 1 halides are a good example (Chapter 4) to strikingly covalent compounds such as CCl_4. In general, the highest possible formal oxidation states of the elements are displayed by the fluorides. Examples include PtF_6 and UF_6. The latter is enormously important. Despite the formal oxidation number of +6, it is highly covalent. It melts at about 64°C and has a vapour pressure of about 115 mm at room temperature. As such, the hexafluoride is used in gas diffusion processes to separate uranium isotopes.

5.11 Group 18 elements

The Group 18 elements are often referred to as *noble gases*. They exist as monoatomic molecules and are found uncombined in nature. The group electronic configuration is ns^2np^6 meaning that the valence shell is closed. While helium ($2s^2$) does not have any p electrons, its valence shell is also closed and most commonly used periodic tables place helium above neon in Group 18. One could make a case for placing helium above beryllium in Group 2 (whose group electron configuration is also ns^2).

The Group 18 gases are trace components in the atmosphere and neon, argon, krypton, and xenon are all isolated through fractionation of liquid air. While helium does occur in the atmosphere, helium is obtained commercially from some natural gas wells, particularly in the USA. Concentrations can be up to 5–7%. It is likely that the origin of the helium in these gases is through the decay of radioactive elements in rocks. All isotopes of radon are radioactive and there is concern in some areas of the world where poor ventilation in homes results in build-up of radon gas (which comes from the rocks upon which the homes are built).

The elements are all colourless. The boiling points of these elements increase smoothly down the group (Fig. 5.22). Helium is unique in that it has no triple point — that is, there is no single combination of temperature and pressure at which the solid, the liquid, and the gas all exist. At around 2 K, helium undergoes a transition to a form of helium labelled He_{II}. This form of helium has strange properties, including the ability to form films only a few hundred atoms thick that flow apparently without friction, even up the sides of a vessel.

The gases are used to fill electric light-bulbs with various interesting colours. In addition they are used to provide inert gases for welding. Argon is used quite commonly as an inert atmosphere under which to carry out sensitive reactions in the laboratory and there is now interest in the use of the liquefied gases as solvents for specialized reactions.

The Group 18 elements have limited reactivity — indeed, the origin of the

Group properties

Largely unreactive monoatomic gases

Group name: noble gases

Group members: (He), Ne, Ar, Kr, Xe, Rn

Group configuration ns^2np^6

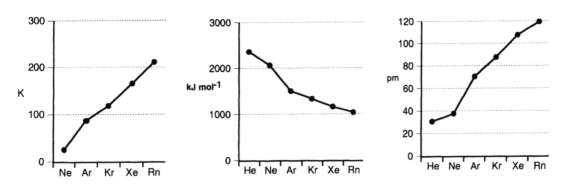

Fig. 5.22 Some group trends of the Group 18 elements: boiling point (left), the first ionization energy (centre,), and atomic radius (right).

term noble gases is linked to their apparent lack of reactivity. What chemistry there is, however, is interesting and largely concerns xenon.

Oxygen is known to react with PtF_6 to form an orange salt $[O_2][PtF_6]$. This material contains discrete O_2^+ cations. The tendency for xenon gas to ionize is about the same as dioxygen and indeed xenon gas does react with PtF_6, although the reaction is complicated.

Most xenon chemistry is associated with the fluorides and oxyfluorides (Fig. 5.23). The difluoride and tetrafluoride are both available through the reaction of xenon with fluorine at high pressure. The hexafluoride is also available directly but requires very forcing conditions.

The shapes of these xenon compounds are successfully predicted by simple VSEPR models. Thus, the σ-framework of XeF_2 is determined by ten electrons, meaning that it is trigonal bipyramidal. On paper a number of isomers are possible but that which minimizes the electron pair–electron pair interactions has the two fluorines in the axial positions, making the molecule linear. The tetrafluoride is square planar. The case of XeF_6 is more interesting. Its shape is determined by seven pairs of electrons. Experimental results indicate that the solid state structure is very complex and distorted.

The fluorides XeF_4 and XeF_6 both react with water to form the explosive trioxide XeO_3. This dangerous property explains why only skilled chemists attempt to work with xenon fluorides, since even trace amounts of water can lead to XeO_3. This compound is pyramidal and the xenon has a directional lone pair. Very careful and controlled reaction of XeF_6 with water results in the oxofluoride XeF_4O (Eqn 5.12). This material reacts with XeO_3 to form the dioxide XeF_2O_2 (Eqn 5.13).

$$XeF_6 + H_2O \rightarrow XeF_4O + 2HF \qquad (5.12)$$

$$XeF_4O + XeO_3 \rightarrow 2XeF_2O_2 \qquad (5.13)$$

Again, the shapes of these two compounds are predicted successfully by applying VSEPR rules. They have shapes dictated by six and five electron pairs respectively. The compound XeF_4O has one lone pair and its shape is therefore square-based pyramidal while the geometry of xenon in XeF_2O_2 is trigonal bipyramidal with equatorial oxide ligands and lone pair.

The hexafluoride XeF_6 is a fluoride ion acceptor. Thus CsF reacts to form the heptafluoroxenate and upon gentle warming this material converts into the remarkably stable octafluoroxenate $Cs_2[XeF_8]$ (Eqn 5.14).

$$2XeF_6 + 2CsF \rightarrow 2Cs[XeF_7] \rightarrow Cs_2[XeF_8] + XeF_6 \qquad (5.14)$$

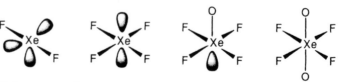

Fig. 5.23 Some xenon fluorides and oxyfluorides.

6 *d*-Block elements

The term *d-block elements* refers to the 30 elements contained in the 10 columns (3–12) in the periodic table (inside back cover). It is also a convenient term since it includes the elements zinc, cadmium, and mercury, some properties of which it is logically appropriate to include in a discussion of transition metal chemistry. The term *transition metal* is normally associated with those elements that possess a *d* sublevel that is partially filled with electrons in either its atom or a common oxidation state. As Table 6.1 shows, by this definition zinc is not a transition element and both scandium and copper just scrape in by virtue of one configuration each.

6.1 Variable oxidation state

Although *d*-block metals are similar in many ways to those of the *s* and *p* blocks they have a much greater tendency to display variable oxidation states. As an example, there are four vanadium fluorides: VF_2 (blue), VF_3 (yellow–green), VF_4 (lime green), and VF_5 (colourless). Table 6.1 shows the electron configuration of each element in the first row of the *d* block together with its common oxidation states.

Figure 6.1 shows that successive ionization energies for metals in the *s* and *p*-blocks generally show a large jump when a noble gas electron configuration is disturbed and this determines the oxidation state of the metal in simple ionic compounds. For a *d*-block metal such as vanadium, however, after the $4s$ electrons have been lost, some, or all, of the $3d$ electrons may usually be removed successively without a sudden jump in ionization energy.

6.2 Transition elements and coloured compounds

Many, but not all, of the compounds of *d*-block elements are unusual in being coloured. Certainly in the case of lower oxidation states this property is often

Table 6.1 Electron configurations of the atoms and common oxidation states of the elements in the first row of the *d* block

Oxidation state		Sc	Ti	V	Cr	Mn	Fe	Co	Ni	Cu	Zn
0	[Ar]	$3d^1 4s^2$	$3d^2 4s^2$	$3d^3 4s^2$	$3d^5 4s^1$	$3d^5 4s^2$	$3d^6 4s^2$	$3d^7 4s^2$	$3d^8 4s^2$	$3d^{10} 4s^1$	$3d^{10} 4s^2$
1	[Ar]									$3d^{10}$	
2	[Ar]		$3d^2$	$3d^3$	$3d^4$	$3d^5$	$3d^6$	$3d^7$	$3d^8$	$3d^9$	$3d^{10}$
3	[Ar]	$3d^0$	$3d^1$	$3d^2$	$3d^3$		$3d^5$	$3d^6$			
4	[Ar]		$3d^0$	$3d^1$							
5	[Ar]			$3d^0$							
6	[Ar]				$3d^0$						
7	[Ar]					$3d^0$					

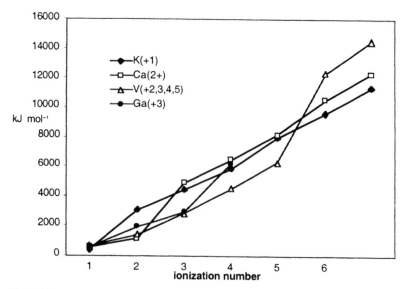

Fig. 6.1 Ionization energies for some elements. Redox reactions involving *d*-block metal species are covered in Section 6.7.

associated with a *d* sublevel that is partially filled with electrons. Recall that any element which possesses this configuration in either its atom or a common oxidation state is a transition metal.

The connection between coloured compounds and a partially filled *d* sublevel can be explored by comparing the energy changes involved in forming the anhydrous chlorides: $CoCl_2$, which is blue ($3d^7$), and $ZnCl_2$, which is

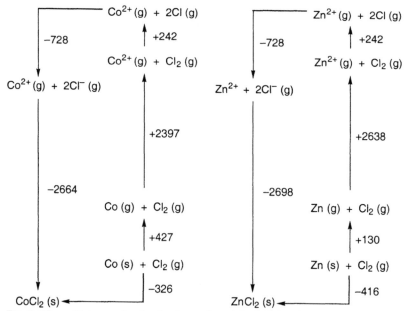

Fig. 6.2 Born–Haber cycles for the formation of the halides $CoCl_2(s)$ and $ZnCl_2(s)$. All values in kJ mol^{-1}.

colourless ($3d^{10}$). There is little indication of any difference from a comparison of the Born–Haber cycles for the formation of the compounds (Fig. 6.2). Each chloride is energetically more stable than its elements thanks largely to a very exothermic lattice energy resulting from the strong attraction between the M^{2+} and Cl^- ions in the crystal. To begin to account for the blue colour of cobalt chloride we must also consider the mutual repulsion of electrons on the metal and halide ions as they approach each other to form the lattice.

If a Co^{2+} or Zn^{2+} ion is in the gas phase, all five of the d orbitals are at exactly the same energy and it does not matter along which directions the x, y, and z axes lie. The five d-orbital functions are arranged in space with respect to each other as shown in Fig. 6.3. There is no energy advantage in an electron occupying any particular orbital. In effect, although the d orbitals are normally labelled according to the cartesian (xyz) coordinate system, the isolated ion has no sense of direction.

However, when the lattice forms, the d electrons on the cation suffer repulsion from the electrons on the surrounding anions. The electrons in the d orbitals now arrange themselves in space so as to minimize this repulsion. This is done so far as possible by the electrons arranging themselves not to point directly towards the anions. They tend to occupy regions of space orientated between the anions. The orientation of the d-orbital system will depend upon the geometry of the surrounding anions in the lattice and in this case the x, y, and z directions are defined so that the anions lie on the coordinate axes.

In the cobalt(II) chloride lattice (Fig. 6.4), each Co^{2+} ion is at the centre of an octahedral arrangement of six Cl^- ions . The d orbitals are arranged so that the d_{z^2} and $d_{x^2-y^2}$ orbitals on the metal ion point directly at the Cl^- ions, with the remaining d_{xy}, d_{xz} and d_{yz} orbitals each pointing between the anions. The d orbitals are now no longer at the same energy, since electrons occupying

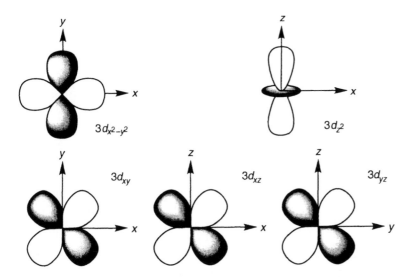

Fig. 6.3 The five d orbitals. Note the changes in axis labelling for each orbital.

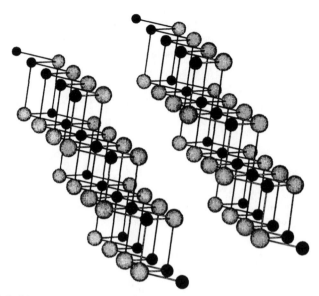

Fig. 6.4 Solid-state structure of CoCl$_2$. Note the layer structure. Each cobalt ion (dark shading) is surrounded by six halides in an octahedral array.

orbitals pointing directly at the chloride ions will suffer greater repulsion than those in the other set of orbitals. In cobalt(II) chloride, the difference in energy between the two sets of *d* orbitals, referred to as the octahedral *crystal field splitting energy*, Δ_o, is less than the energy required to overcome the repulsion between electrons occupying the same orbital, so the ground state electronic arrangement of the *d* electrons will be as shown in Fig. 6.5.

Cobalt(II) chloride absorbs visible light towards the red end of the spectrum. Light at that wavelength possesses the correct energy to promote an electron from the *lower* to the *higher* energy set of *d* orbitals. Light from the blue end of the visible spectrum is not absorbed and it is this unabsorbed

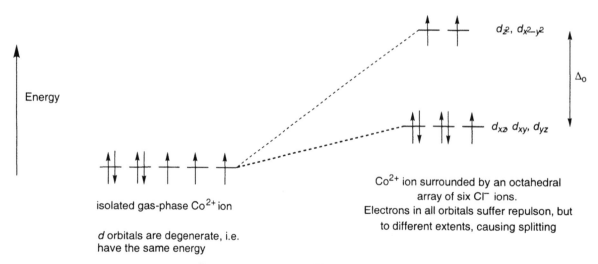

Fig. 6.5. Effect on the 3d^7 arrangement of the Co^{2+} ion on forming the CoCl$_2$ crystal lattice.

light that accounts for the observed colour of cobalt(II) chloride.

The *d* orbitals in zinc chloride also suffer crystal field splitting. However, since all the orbitals are fully occupied, no promotion from the lower set of *d* orbitals to the higher is possible and so this compound does not absorb visible light.

6.3   Aqua complexes: dissolving a metal compound in water

Both cobalt(II) chloride and zinc chloride dissolve exothermically in water. Since water molecules are polar, they are strongly attracted to the ions and the combined energy released by these hydration processes is more than sufficient to overcome the attraction between the ions in the lattice (Fig. 6.6).

As might be expected from the previous section, zinc chloride solution is colourless. However, although dilute aqueous cobalt(II) chloride is coloured, it is *pink* and not *blue* like the anhydrous solid. The reason for this lies in the change in the environment of the Co^{2+} ions upon dissolving. The chloride ions surrounding the cobalt ion in the solid lattice are replaced by water molecules.

The arrangement of attached atoms around the Co(II) is still octahedral but the water molecules produce a larger crystal field splitting energy so absorbing light of higher quantum energy towards the blue end of the spectrum. Light from the red end of the visible spectrum is not absorbed, leading to the observed pink colour of the solution.

All *d*-block metal ions are hydrated in aqueous solution but usually the resulting species will not be coloured unless it possesses a partially filled *d* sublevel. The central metal ion will attract a lone pair of electrons from each of the water molecules to some extent and so the bonding between them may be regarded as somewhere between simple electrostatic attraction and coordinate covalent (Fig. 6.7).

Species like the hydrated cobalt(II) ion that contain a central cation surrounded by electron pair donors are referred to as *complexes*. The electron pair donors are called *ligands* and the number of coordinate bonds formed is the *coordination number*. When writing the formula of a complex, it is

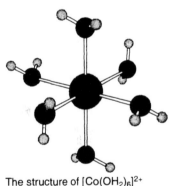

The structure of $[Co(OH_2)_6]^{2+}$

Electron pair donors are also referred to as *Lewis bases* and are nucleophilic.

Co^{2+} (g) + 2Cl⁻ (g)

$\downarrow$ −728

+2664

Co^{2+} (g) + 2Cl⁻ (aq)

$\downarrow$ −2023

CoCl$_2$ (s)

−87

Co^{2+} (aq) + 2Cl⁻ (g)

Zn^{2+} (g) + 2Cl⁻ (g)

$\downarrow$ −728

+2698

Zn^{2+} (g) + 2Cl⁻ (aq)

$\downarrow$ −2042

ZnCl$_2$ (s)

−72

Zn^{2+} (aq) + 2Cl⁻ (g)

Fig. 6.6 Enthalpy changes involved in dissolving CoCl$_2$(s) and ZnCl$_2$(s) in excess water. All values in kJ mol^{-1}.

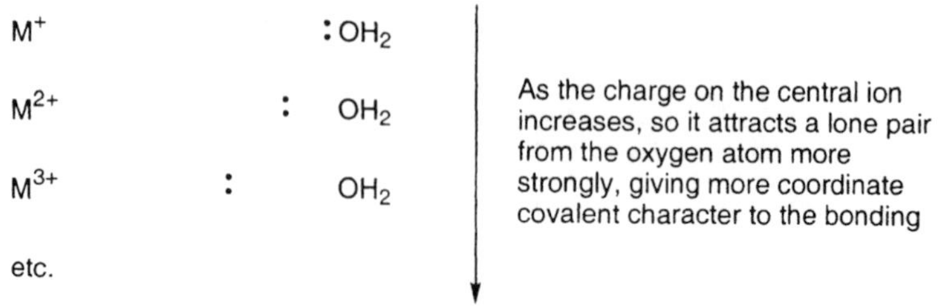

M$^+$:OH$_2$

M^{2+} :   OH$_2$

M^{3+} :   OH$_2$

As the charge on the central ion increases, so it attracts a lone pair from the oxygen atom more strongly, giving more coordinate covalent character to the bonding

etc.

Fig. 6.7 The increase in covalency with increase in charge on the metal ion.

conventional to enclose it in square brackets, for example: $[Co(H_2O)_6]^{2+}$.

All metals in the first row of the d-block with an oxidation state of +2 or +3 form such hexa-aqua complexes in aqueous solution. In many cases, the solids that may be crystallized from aqueous solution retain the hydrated metal ion. This includes the result of crystallizing pink hydrated cobalt(II) chloride. It is common to see these denoted as $CoCl_2 \cdot 6H_2O$, but a more informative representation of the formula is $[Co(H_2O)_6]Cl_2$.

6.4   Acidity of aqua complexes

All hydrated cations are acidic to some extent in solution. Electron density is attracted from coordinated water molecules increasing the partial positive charge on the hydrogen atoms and making them more susceptible to attack by other water molecules to produce $H_3O^+(aq)$ (Fig. 6.8).

This tendency increases with the charge density on the cation. So, $M^{3+}(aq)$ is more acidic than $M^{2+}(aq)$.

Such hydrolysis reactions cannot occur in the solid state and this may

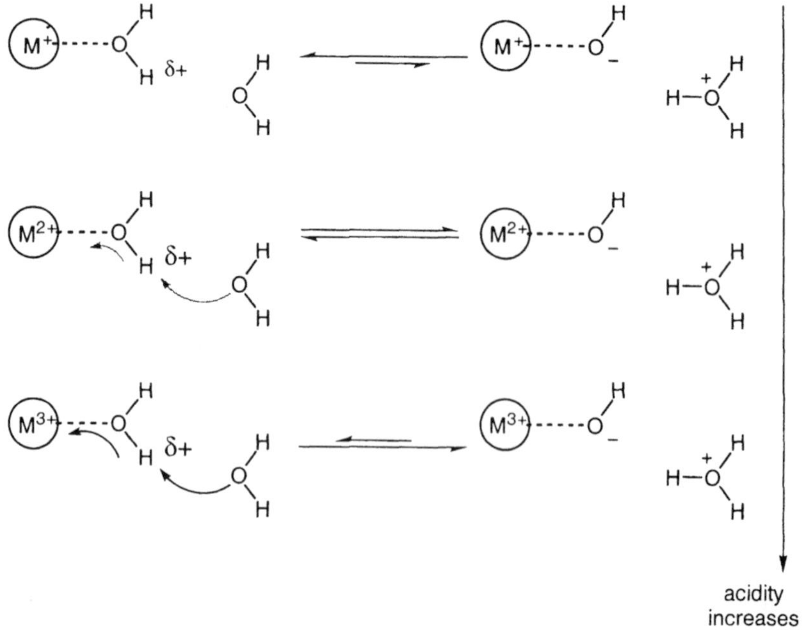

acidity increases

Fig. 6.8 The increase in acidity in metal aqua complexes with increasing charge.

$$\left[\begin{array}{c} O \\ \| \\ O-Mn^{VII} \\ \diagup \diagdown \\ O \quad O \end{array}\right]^{-} \qquad \left[\begin{array}{c} O \\ \| \\ O-Cr^{VI} \\ \diagup \diagdown \\ O \quad O \end{array}\right]^{2-} \qquad \left[\begin{array}{c} O \quad O \\ \| \quad \| \\ O-Cr^{VI} \quad Cr^{VI}-O \\ \diagup \diagdown O \diagdown \\ O \quad O \end{array}\right]^{2-}$$

Fig. 6.9 Complexes with colour due to charge transfer from ligand to metal.

explain why some $[M(H_2O)_6]^{3+}$ ions show different colours in the solid state from that in aqueous solution. As an example, pink crystals of $Fe(NO_3)_3.9H_2O$ dissolve in water to give a yellow–brown solution.

When the oxidation state of the metal ion is higher than $+3$, the attraction for electrons on coordinated water molecules is so strong that *both* the hydrogen atoms may be lost. Thus, manganese(VII) in aqueous solution forms manganate(VII), $[MnO_4]^-$ (or permanganate). Although, nominally, the manganese in manganese(VII) has no d electrons, the complex $[MnO_4]^-$ is intensely purple. In this case, absorption of visible light causes electrons to jump from filled orbitals on the oxo ligands into empty orbitals on the central metal. Such *charge-transfer* mechanisms are also responsible for the colour of other species containing d-block metals in a high oxidation state, including chromate(VI) and dichromate(VI) (Fig. 6.9).

Hexa-aqua complexes also undergo characteristic reactions with other bases. As an example, addition of sodium hydroxide solution to $[M(OH_2)_6]^{2+/3+}(aq)$ always first produces a precipitate. Formation of a precipitate in this way is a good indicator of the formation of an electrically neutral species. The precipitate is formed by losing an H^+ from the required number of coordinated water molecules (Eqns 6.1 and 6.2).

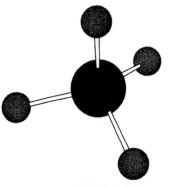

The structure of the $[MnO_4]^-$ ion.

$$[M(OH_2)_6]^{2+}(aq) + 2OH^-(aq) \rightleftharpoons [M(OH)_2(OH_2)_4](s) + 2H_2O(l) \qquad (6.1)$$

$$[M(OH_2)_6]^{3+}(aq) + 3OH^-(aq) \rightleftharpoons [M(OH)_3(OH_2)_3](s) + 3H_2O(l) \qquad (6.2)$$

Addition of acid shifts the equilibrium to the left, and the precipitate, acting as a base, accepts H^+. It redissolves to form the hexa-aqua complex.

In a few cases, including Cr(III) and Zn(II), addition of excess sodium hydroxide solution also dissolves the initial precipitate by removing further H^+ ions to form an *anionic* complex (Eqn 6.3). Substances like $[Cr(OH)_3(H_2O)_3]$ that can both accept and donate H^+ are referred to as *ambiprotic* (or *amphoteric*).

$$[Cr(OH)_3(OH_2)_3](s) + 3OH^-(aq) \rightleftharpoons [Cr(OH)_6]^{3-}(aq) + 3H_2O(l) \qquad (6.3)$$

Carbonate ions, CO_3^{2-}, also accept H^+ to form water and carbon dioxide (Eqn 6.4). However, CO_3^{2-} is a much weaker base than OH^- and only undergoes this reaction with moderately strong acids. Addition of sodium carbonate solution therefore gives a precipitate of the neutral hydroxo complex with $M^{3+}(aq)$ that is insoluble in excess of the reagent (Eqn 6.5).

$$CO_3^{2-}(aq) + 2H^+(aq) \rightarrow H_2O(l) + CO_2(g)$$

(6.4)

$$2[M(OH_2)_6]^{3+}(aq) + 3CO_3^{2-}(aq) \rightarrow$$
$$2[M(OH)_3(OH_2)_3](s) + 3H_2O(l) + 3CO_2(g)$$

(6.5)

Although no gas is evolved when sodium carbonate solution is added to $M^{2+}(aq)$, a precipitate is formed. This is the metal carbonate rather than the neutral hydroxo complex (Eqn 6.6).

$$M^{2+}(aq) + CO_3^{2-}(aq) \rightarrow MCO_3(s)$$

(6.6)

Initially, at least, dilute ammonia solution reacts in exactly the same way as sodium hydroxide, giving a precipitate of the electrically neutral hydroxide with both $M^{2+}(aq)$ and $M^{3+}(aq)$ (Eqns 6.7 and 6.8).

$$[M(OH_2)_6]^{2+}(aq) + 2NH_3(aq) \rightarrow [M(OH)_2(OH_2)_4](s) + 2NH_4^+(aq)$$

(6.7)

$$[M(OH_2)_6]^{3+}(aq) + 3NH_3(aq) \rightarrow [M(OH)_3(OH_2)_3](s) + 3NH_4^+(aq)$$

(6.8)

Although aqueous ammonia is not a sufficiently strong base to remove further protons from the neutral hydroxo complex, the precipitate nonetheless often dissolves on adding an excess of the reagent. This is because as well as acting as a base, ammonia can also compete with water as a ligand (by donating the lone pair of electrons from the nitrogen atom to the central metal) resulting in the formation of soluble charged complexes.

6.5 Ligand exchange reactions involving aqua complexes

Neutral hydroxo complexes often dissolve in excess aqueous ammonia owing to ligand exchange producing cationic ammine complexes. Since water and ammonia molecules are roughly the same size, there is no change in coordination number (Eqn 6.9). In the case of copper(II), aqueous ammonia does not replace all the water molecules (Eqn 6.10).

$$[Cr(OH)_3(H_2O)_3](s) + 6NH_3(aq) \rightleftharpoons$$
$$[Cr(NH_3)_6]^{3+}(aq) + 3H_2O(l) + 3OH^-(aq)$$

(6.9)

$$[Cu(OH)_2(H_2O)_4](s) + 4NH_3(aq) \rightleftharpoons$$
$$[Cu(H_2O)_2(NH_3)_4]^{2+}(aq) + 2H_2O(l) + 2OH^-(aq)$$

(6.10)

Neither $[Fe(OH)_2(H_2O)_4]$ nor $[Fe(OH)_3(H_2O)_3]$ dissolves in excess aqueous ammonia. Apparently, iron does not form stable ammine complexes under these conditions.

Ligand exchange is nearly always accompanied by a *colour* change. In the case of ammine formation there is no change in the *geometry* of the complex but ammonia ligands are generally rather better at splitting the *d* orbitals on

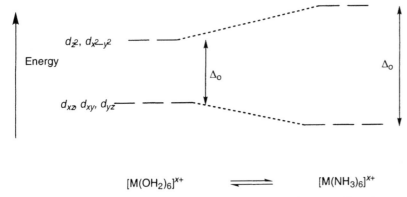

$$[M(OH_2)_6]^{x+} \quad \rightleftharpoons \quad [M(NH_3)_6]^{x+}$$

Fig. 6.10 Ammine ligands cause a greater *d*-orbital splitting than aqua ligands.

the metal than aqua ligands. This means that the *d–d* absorption bands will occur at higher energy, that is, towards the blue–violet region (Fig. 6.10).

The coordinated water molecules in aqua complexes may also undergo exchange with other ligands. Addition of concentrated hydrochloric acid, for example, often results in a *tetrahedral* chloro complex (Eqn 6.11).

$$[Co(H_2O)_6]^{2+}(aq) + 4Cl^-(aq) \rightleftharpoons [CoCl_4]^{2-}(aq) + 6H_2O(l) \qquad (6.11)$$

The change in coordination number here may be explained in terms of the relative sizes of the competing ligands. Since chloride ions are larger than water molecules, fewer can fit around the central metal ion. Since the

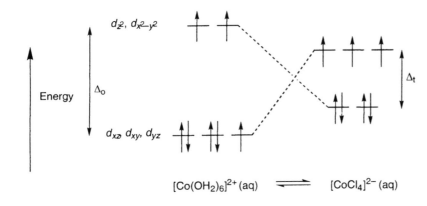

$$[Co(OH_2)_6]^{2+}(aq) \quad \rightleftharpoons \quad [CoCl_4]^{2-}(aq)$$

Fig. 6.11 *d*-Orbital splittings for octahedral (left) and tetrahedral (right) Co(II) fields.

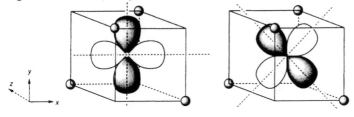

Fig. 6.12 The $d_{x^2-y^2}$ orbital (left) and the d_{xy} orbital (right) in a tetrahedral field.

geometry of the complex is changed, the *d*-orbital splitting pattern, which is caused by interactions with the ligand electrons, is also changed. In the case of a tetrahedral crystal field, minimum electron repulsion is achieved by the *d* orbital orientation shown in Fig. 6.11. Neither of the two sets of *d* orbitals

---

Naming complexes

There are systematic rules for naming coordination compounds. Some of the important conventions are shown below.

1   Cationic or anionic species in complexes have **single word** names in which the ligands precede the metal ion.

2a   Negatively charged ligands are named before neutral ligands, and within each group, ligands are listed in alphabetical order.

2b   Names of negative ion ligands end in '*o*', so Cl^- is *chloro*. Neutral ligands often keep normal names. Exceptions include the common ligands H_2O, which is called *aqua*, and NH_3, which is called *ammine*.

2c   The number of each type of ligand is indicated by the prefixes:

1	*mono-*
2	*di-*
3	*tri-*
4	*tetra-*
5	*penta-*
6	*hexa-*

3   If the complex has an overall negative charge the metal name is given the ending *-ate*. The Latin form is sometimes used in this case, iron (ferrate), copper (cuprate), and silver (argentate). The oxidation state of the metal is written in Roman numerals, in brackets, after its name.

4   In ionic compounds, the positive ion is named *before* the negative ion regardless of whether either is a complex.

Although these rules may seem complicated, the following examples will show that, in practice, they are quite simple to apply.

$[Ni(OH_2)_6]Cl_2$ This contains $[Ni(H_2O)_6]^{2+}$ and $2Cl^-$. The cation is named first followed by the anion: hexaaquanickel(II) chloride. Note the space between the two ions and that there is no need to state the number of chloride ions.

$[Fe(OH)_3(OH_2)_3]$ This is a neutral complex and so has a single word name: triaquatrihydroxoiron(III). Note that the ligands are listed alphabetically, disregarding the prefix.

$Na_2[CoCl_4]$ This contains $2Na^+$ and $[CoCl_4]^{2-}$. The cation is named first followed by the complex anion: sodium tetrachlorocobaltate(II). Note the '*ate*' ending of negatively charged complex ions.

points directly at the ligands (Fig. 6.12), so the splitting energy for a tetrahedral complex, Δ_t, is less than that for an octahedral complex, Δ_o.

The colour change from pink to blue results from a shift in absorption towards the lower energy, that is, the red region of the spectrum. This shift is due partly to the change in geometry of the complex and partly to the weaker splitting power of chloride ligands.

6.6 Types of ligand

Several of the simple ligands mentioned in previous sections possess more than one lone pair of electrons (Fig. 6.13). Although such species may act as bridges by donating an electron pair to each of two metal ions as in the dichromate(VI) ion (Fig. 6.9), these ligands are referred to as monodentate as only one atom can donate electron pairs.

Fig. 6.13 Some ligands with more than one lone pair.

Some ligands, however, contain two or more atoms that can act as electron pair donors. Examples of such didentate, tridentate, and tetradentate ligands are shown in Fig. 6.14. A metal ion that accepts two or more pairs of electrons from the same ligand becomes incorporated into a ring system known as a chelate. Such complexes are generally much less susceptible to exchange reactions than non-chelates since two or more coordinate bonds must break almost simultaneously in order to release the ligand. Thus the EDTA anion, which can supply up to six electron pairs to a single metal ion, forms extremely stable complexes of the type shown in Fig. 6.15.

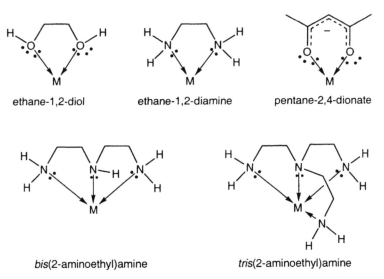

ethane-1,2-diol   ethane-1,2-diamine   pentane-2,4-dionate

bis(2-aminoethyl)amine   *tris*(2-aminoethyl)amine

Fig. 6.14 Some multidentate ligands

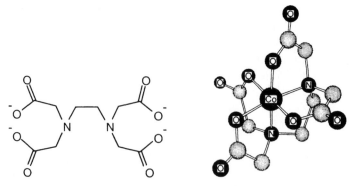

Fig. 6.15 The structure of *bis*[di(carboxymethyl)-amino]ethane tetraanion (commonly known as $EDTA^{4-}$) and the structure of an EDTA cobalt complex.

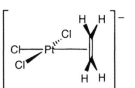

Fig. 6.16 The ion $[PtCl_3(C_2H_4)]^-$.

Although they do not possess any lone pairs, alkenes act as ligands by donating a pair of π-bonding electrons, as in the ion $[PtCl_3(C_2H_4)]^-$ (Fig. 6.16). Molecules such as benzene that contain a delocalized ring of π electrons can act as ligands, for example in the 'sandwich' molecules dibenzenechromium, $[Cr(C_6H_6)_2]$, and ferrocene (Fig. 6.17).

Fig. 6.17 The sandwich molecules ferrocene (left) and dibenzenechromium (right).

6.7 Standard potentials, $E°$, and redox reactions

This section explains the term *standard electrode potentials* and indicates how they may be used to predict the course of redox reactions. Redox reactions involve the transfer of electrons. Any species that loses electrons is said to be oxidized whilst those that gain electrons are reduced. Neither of these processes can occur in isolation and must occur simultaneously.

When a metal is dipped into a solution of its ions, a redox equilibrium is set up (Eqn 6.12).

$$M^{x+}(aq) \rightleftharpoons M(s) \qquad (6.12)$$

The better the metal (M) is at giving away electrons, the more negative will be the charge that the solid metal develops at equilibrium.

It is not possible to measure the potential of a single redox electrode in isolation but it is possible to measure the potential *difference* between any two electrodes by setting up a chemical cell. Like any other equilibrium, the position of a redox system will depend upon the conditions chosen. *The standard potential of any system is defined as its potential measured relative to the hydrogen electrode (Eqn 6.13) at 25°C, 1 atmosphere pressure, and with all aqueous concentrations equal to 1 M.*

the hydrogen electrode system: $H^+(aq) + e^- \rightleftharpoons 0.5H_2(g)$ (6.13)

The more *negative* the value of the standard electrode potential, the more easily the system loses electrons so the *greater* is its *reducing power* (and the weaker its oxidizing power.) We can, therefore, use E° values to predict the course of reactions involving redox equilibria.

For example, consider the addition of magnesium metal to dilute sulphuric acid. The two redox systems involved are shown in Eqns 6.14 and 6.15.

$$Mg^{2+}(aq) + 2e^- \rightleftharpoons Mg(s) \qquad E^\circ = -2.38 \text{ V} \qquad\qquad (6.14)$$

$$H^+(aq) + e^- \rightleftharpoons 0.5H_2(g) \qquad E^\circ = 0.00 \text{ V} \qquad\qquad (6.15)$$

When the two systems are in contact, electrons will flow if possible from the more negative system (Eqn 6.14) to the more positive (Eqn. 6.15).

$$Mg^{2+}(aq) + 2e^- \rightleftharpoons Mg(s)\ E^\circ = -2.38 \text{ V}$$

electron flow

$$H^+(aq) + e^- \rightleftharpoons 0.5H_2(g)\ E^\circ = 0.00 \text{ V}$$

This will disturb the above redox equilibria, and reactions will take place producing aqueous magnesium ions and hydrogen gas:

$$Mg(s) \rightarrow Mg^{2+}(aq) + 2e^-$$

$$H^+(aq) + e^- \rightarrow 0.5H_2(g)$$

To get the overall equation it is necessary to ensure that the number of electrons released by the reducing agent equals the number accepted by the oxidizing agent. In this case this balance is achieved by doubling the values in Eqn 6.15.

$$Mg(s) \rightarrow Mg^{2+}(aq) + 2e^-$$

$$2H^+(aq) + 2e^- \rightarrow H_2(g)$$

overall equation $= Mg(s) + 2H^+(aq) \rightarrow Mg^{2+}(aq) + H_2(g)$

So, in this case a reaction is predicted between magnesium and dilute sulphuric acid, although strictly speaking this is only valid for the conditions

LIVERPOOL JOHN MOORES UNIVERSITY
LEARNING SERVICES

at which standard electrode potentials are measured. Now consider adding copper metal to dilute sulphuric acid. The two reactions in question are given in Eqns 6.16 and 6.17.

$$Cu^{2+}(aq) + 2e^- \rightleftharpoons Cu(s) \quad E° = +0.34 \text{ V} \qquad (6.16)$$

$$H^+(aq) + e^- \rightleftharpoons 0.5H_2(g) \quad E° = 0.00 \text{ V} \qquad (6.17)$$

In this case, a flow of electrons from the more negative hydrogen system to the copper system is predicted.

$$Cu^{2+}(aq) + 2e^- \rightleftharpoons Cu(s) \quad E° = +0.34 \text{ V}$$

electron flow

$$H^+(aq) + e^- \rightleftharpoons 0.5H_2(g) \quad E° = 0.00 \text{ V}$$

The predicted reactions are given below and the overall reaction is their sum. Copper metal will not, therefore react with $H^+(aq)$ under standard conditions. In fact, the reverse process, that is the reduction of Cu^{2+} by hydrogen, is predicted.

$$Cu^{2+}(aq) + 2e^- \rightarrow Cu(s)$$

$$H_2(g) \rightarrow 2H^+(aq) + 2e^-$$

overall equation $= Cu^{2+}(aq) + H_2(g) \rightarrow Cu(s) + 2H^+(aq)$

6.8 Redox reactions and catalytic action

Higher oxidation states of transition elements are often reduced in acid solution by reducing agents such as zinc metal. For example, when zinc is added to a solution containing ferric ion, $Fe^{3+}(aq)$, the result is reduction to ferrous ion, $Fe^{2+}(aq)$ (Fig. 6.18). The colour change in this reaction results from a change in the number of electrons populating the *d* orbitals. As with ligand exchange, characteristic colour changes in redox reactions can be used in chemical analysis.

If the metal is capable of displaying multiple oxidation states, then a series of colour changes may well indicate progressive reduction. For vanadate(V) the following sequence is observed upon reduction with zinc metal:

$$Zn^{2+}(aq) + 2e^- \rightleftharpoons Zn(s) \qquad E^\circ = -0.76 \text{ V}$$

electron transfer

$$Fe^{3+}(aq) + e^- \rightleftharpoons Fe^{2+}(aq) \qquad E^\circ = +0.77 \text{ V}$$

reduction	$2Fe^{3+}(aq) + 2e^-$	$\rightarrow \quad 2Fe^{2+}(aq)$
oxidation	$Zn(s)$	$\rightarrow \quad Zn^{2+}(aq) + 2e^-$

redox	$2Fe^{3+}(aq) + Zn(s)$	$\rightarrow \quad 2Fe^{2+}(aq) + Zn^{2+}(aq)$
	yellow-brown	pale green

Fig. 6.18 The reduction of $Fe^{3+}(aq)$ to $Fe^{2+}(aq)$ by zinc.

	$VO_3^-(aq) \rightarrow$	$VO^{2+}(aq) \rightarrow$	$V^{3+}(aq) \rightarrow$	$V^{2+}(aq)$
oxidation number	+5	+4	+3	+2
	pale yellow	bright blue	green	violet

If the resulting final solution is separated from the zinc and allowed to stand in air it reverts to a bright blue colour, showing that +4 is the most stable oxidation state of vanadium under these conditions. Reduction of dichromate(VI) solution by zinc takes place as follows:

	$Cr_2O_7^{2-}(aq) \rightarrow$	$Cr^{3+}(aq) \rightarrow$	$Cr^{2+}(aq)$
oxidation number	+6	+3	+2
	orange	green	sky-blue
		(most stable in air)	(readily oxidized by air)

Intermediate oxidation states are not always observed. For instance, in strongly acidic solution manganate(VII) is reduced directly to manganese(II):

	$MnO_4^-(aq) \rightarrow$	$Mn^{2+}(aq)$
oxidation number	+7	+2
	deep purple	almost colourless (most stable)

Although the 'most stable' oxidation states are labelled in the above sequences, it should be noted that this can depend upon the conditions or the type of ligand. Thus $[Co(H_2O)_6]^{2+}$ is perfectly stable in air, whereas $[Co(NH_3)_6]^{2+}$ is readily oxidized under the same conditions (Fig. 6.19).

Disproportionation involves self-oxidation and reduction. For example, copper(I) oxide dissolves in dilute acid to give copper metal and $Cu^{2+}(aq)$ (Fig. 6.20). Silver(I), however, is stable in aqueous solution (Fig. 6.21). Redox reactions are also involved when transition metal compounds act as

Disproportionation: self-oxidation and reduction.

$$[Co(OH_2)_6]^{3+} (aq) + e^- \rightleftharpoons [Co(OH_2)_6]^{2+} (aq) \qquad E^° = +1.82 \text{ V}$$

electron transfer ↑

$$\tfrac{1}{2}O_2 (g) + 2H^+ (aq)\ 2e^- \rightleftharpoons H_2O (l) \qquad E^° = +1.23 \text{ V}$$

electron transfer ↑

$$[Co(NH_3)_6]^{3+} (aq) + e^- \rightleftharpoons [Co(NH_3)_6]^{2+} (aq) \qquad E^° = -0.43 \text{ V}$$

readily oxidized

Fig. 6.19 The relative stabilities of $[Co(H_2O)_6]^{2+}$ and $[Co(NH_3)_6]^{2+}$ in air.

$$Cu^+ (aq) + e^- \rightleftharpoons Cu (s) \qquad E^° = +0.52 \text{ V}$$

electron transfer ↑

$$Cu^{2+} (aq) + e^- \rightleftharpoons Cu^+ (aq) \qquad E^° = +0.15 \text{ V}$$

reduction	$Cu^+ (aq) + e^-$	$\rightarrow \quad Cu (s)$
oxidation	$Cu^+ (aq)$	$\rightarrow \quad Cu^{2+} (aq) + e^-$
redox	$2Cu^+ (aq)$	$\rightarrow \quad 2Cu (s) + Cu^{2+} (aq)$

Fig. 6.20 The disproportionation of Cu(I).

$$Ag^+ (aq) + e^- \rightleftharpoons Ag (s) \qquad E^° = +0.80 \text{ V}$$

electron transfer ↓

$$Ag^{2+} (aq) + e^- \rightleftharpoons Ag^+ (aq) \qquad E^° = +1.98 \text{ V}$$

reduction	$Ag^{2+} (aq) + e^-$	$\rightarrow \quad Ag^+ (aq)$
oxidation	$Ag (s)$	$\rightarrow \quad Ag^+ (aq) + e^-$
redox	$Ag^{2+} (aq) + Ag (s)$	$\rightarrow \quad 2Ag^+ (aq)$

Fig. 6.21 Disproportionation of Ag(I) does not occur.

catalysts. For example, iodide is oxidized to iodine by persulphate in aqueous solution (Fig. 6.22).

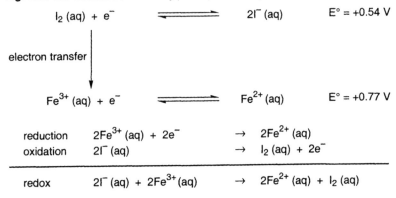

$$I_2 \text{ (aq)} + 2e^- \rightleftharpoons 2I^- \text{ (aq)} \qquad E° = +0.54 \text{ V}$$

electron transfer

$$S_2O_8{}^{2-}\text{(aq)} + 2e^- \rightleftharpoons 2SO_4{}^{2-} \text{ (aq)} \qquad E° = +2.01 \text{ V}$$

reduction	$S_2O_8{}^{2-}\text{(aq)} + 2e^-$	$\rightarrow$	$2SO_4{}^{2-} \text{ (aq)}$
oxidation	$2I^- \text{ (aq)}$	$\rightarrow$	$I_2 \text{ (aq)} + 2e^-$
redox	$S_2O_8{}^{2-}\text{(aq)} + 2I^- \text{ (aq)}$	$\rightarrow$	$SO_4{}^{2-}\text{(aq)} + I_2 \text{ (aq)}$

Fig. 6.22 The oxidation of iodide by persulphate in aqueous solution.

$$I_2 \text{ (aq)} + e^- \rightleftharpoons 2I^- \text{ (aq)} \qquad E° = +0.54 \text{ V}$$

electron transfer

$$Fe^{3+} \text{ (aq)} + e^- \rightleftharpoons Fe^{2+} \text{ (aq)} \qquad E° = +0.77 \text{ V}$$

reduction	$2Fe^{3+} \text{ (aq)} + 2e^-$	$\rightarrow$	$2Fe^{2+} \text{ (aq)}$
oxidation	$2I^- \text{ (aq)}$	$\rightarrow$	$I_2 \text{ (aq)} + 2e^-$
redox	$2I^- \text{ (aq)} + 2Fe^{3+}\text{(aq)}$	$\rightarrow$	$2Fe^{2+} \text{ (aq)} + I_2 \text{ (aq)}$

Fig. 6.23 The oxidation of iodide by iron(III) in aqueous solution.

$$Fe^{3+} \text{ (aq)} + e^- \rightleftharpoons Fe^{2+} \text{ (aq)} \qquad E° = +0.77 \text{ V}$$

electron transfer

$$S_2O_8{}^{2-}\text{(aq)} + 2e^- \rightleftharpoons 2SO_4{}^{2-} \text{ (aq)} \qquad E° = +2.01 \text{ V}$$

reduction	$S_2O_8{}^{2-}\text{(aq)} + 2e^-$	$\rightarrow$	$2SO_4{}^{2-} \text{ (aq)}$
oxidation	$2Fe^{2+} \text{ (aq)}$	$\rightarrow$	$2Fe^{3+} \text{ (aq)} + 2e^-$
redox	$S_2O_8{}^{2-}\text{(aq)} + 2Fe^{2+} \text{ (aq)}$	$\rightarrow$	$SO_4{}^{2-}\text{(aq)} + 2Fe^{3+}\text{(aq)}$

Fig. 6.24 The reduction of persulphate by iron(II) in aqueous solution.

Although very favourable in terms of electrode potentials, the reaction is very slow since the negative charge on each species will result in mutual repulsion as they approach each other. Addition of a small amount of Fe^{3+}(aq), however, speeds up the reaction dramatically by providing an alternative reaction pathway. The first step involves oxidation of iodide by iron(III) (Fig 6.23). The persulphate is then reduced by the iron(II) formed (Fig 6.24).

The iron has acted as an intermediary in the transfer of an electron from I^- to $S_2O_8^{2-}$. Since it both speeds up the reaction and is reformed, the Fe^{3+}(aq) has acted as a catalyst. Of course, Fe^{2+}(aq) would be equally effective.

Before leaving the topic of oxidation and reduction, consider the preparation from iron of the anhydrous chlorides, $FeCl_2$ and $FeCl_3$. Again the conditions are important in deciding which product is formed. If iron is heated in a stream of chlorine, the oxidizing nature of the halogen means that the higher chloride is produced (Eqn 6.18). This can be reduced to iron(II) chloride by heating in a stream of hydrogen (Eqn 6.19).

$$2Fe\ (s) + 3Cl_2\ (g) \rightarrow 2FeCl_3\ (s) \tag{6.18}$$

$$2FeCl_3\ (s) + H_2(g) \rightarrow 2FeCl_2\ (s) + 2HCl\ (g) \tag{6.19}$$

Iron(II) chloride can be prepared directly from the metal by heating in a stream of hydrogen chloride gas. The hydrogen that is also produced ensures that the lower oxidation state is formed (Eqn 6.20).

$$Fe\ (s) + 2HCl\ (g) \rightarrow FeCl_2\ (s) + H_2\ (g) \tag{6.20}$$

6.9 Some important *d*-block compounds

In the space available here it is only possible to give a flavour of the importance and diverse uses of *d*-block compounds. It is true to say that animal life itself depends upon transition metal complexes such as *haemoglobin*. This molecule, which contains Fe^{2+} at the centre of a tetradentate haem group, acts as the oxygen carrier in blood by accepting a lone pair of electrons from O_2 (Fig. 6.25). Carbon monoxide is poisonous because it binds more strongly to the iron that the oxygen molecule. Patients suffering from certain types of anaemia are treated with tablets containing iron(II) sulphate which boosts the iron levels in the blood.

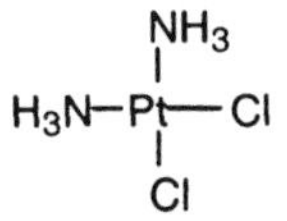

One of the most successful anti-cancer drugs is not a complicated organic molecule but a square planar complex of platinum. The discovery of the anti-cancer properties of cisplatin was completely unexpected and arose from research into the effect of electric fields on solutions containing live bacteria.

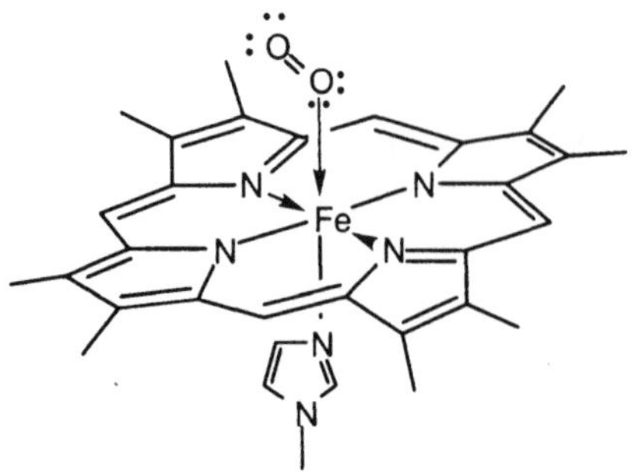

cisplatin

Fig. 6.25 The haemoglobin environment showing the coordinated dioxygen.

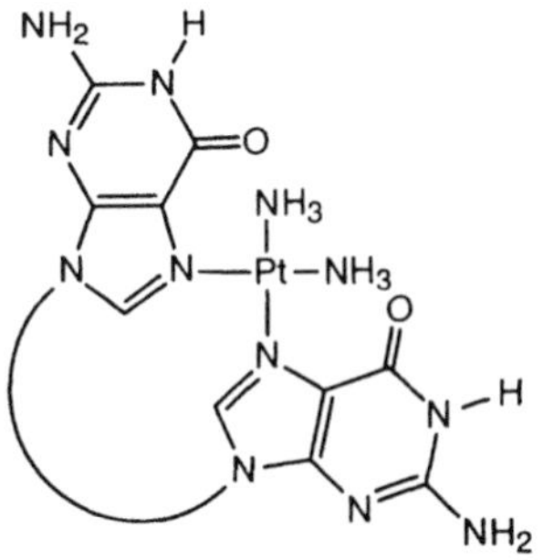

Fig. 6.26 The binding of platinum to a segment of the biological molecule DNA.

only the *cis* isomer has the correct shape to attach itself to adjacent guanines on the DNA chain in this way

Platinum was chosen for the electrodes as it is a very unreactive metal and was therefore unlikely to influence the results.

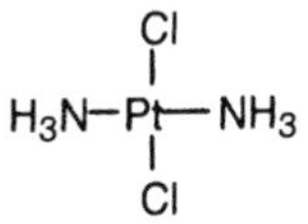

It was found that, although individual cells continued to grow normally, the formation of new cells by division was inhibited. After exhausting all other possible variables in the experiment that might have caused this effect, the researchers concluded that perhaps the platinum had reacted with the solution and produced a new chemical that interfered with cell division. They found very tiny quantities (about 10 parts per million) of cisplatin, together with its isomer transplatin (chloride ligands opposite each other rather than side by side). Further research showed that, unlike cisplatin, transplatin has no effect on cell division.

Since tumours result from uncontrolled division of cancer cells, it was reasoned that cisplatin might be an effective treatment if injected directly into a tumour. This was confirmed by trials and the drug is now a common and effective treatment for testicular cancer provided it is diagnosed early enough.

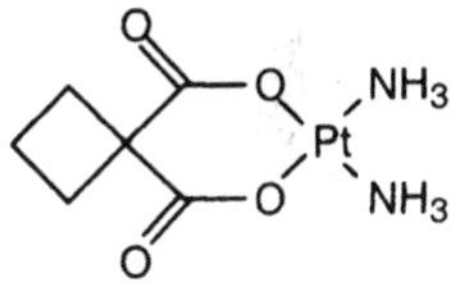

Although very effective, cisplatin is rather toxic and has unpleasant side-effects. This provoked research into the mechanism of its anti-cancer properties so that safer alternatives might be developed. It appears that cisplatin loses its two chloride ligands and then binds to a specific site (the N–7 atoms of adjacent guanines) on the DNA chain (Fig. 6.26). Over 2000 related compounds have undergone medical trials and carboplatin is now in clinical use. This retains the *cis* arrangement of the NH_3 molecules in cisplatin but replaces the chloride ligands with an organic group. It is almost as effective as cisplatin but much less toxic.

In the last section the importance of variable oxidation state in the catalytic action of transition metal compounds was explained. In the contact process for the manufacture of sulphuric acid, vanadium(V) oxide is used as a heterogeneous catalyst for the oxidation of sulphur dioxide to sulphur trioxide. In the uncatalysed reaction (Eqn 6.21), sulphur dioxide is oxidized to sulphur trioxide. In the catalysed reaction, vanadium pentoxide gives up an oxygen to sulphur dioxide (Eqn 6.22) before being reoxidized by oxygen (Eqn 6.23).

$$2SO_2(g) + O_2(g) \rightarrow 2SO_3(g) \tag{6.21}$$

transplatin

carboplatin

$$2SO_2 (g) + 2V_2O_5 (s) \rightarrow 2SO_3 (g) + 2V_2O_4 (s)$$

(6.22)

$$2V_2O_4 (s) + O_2 (g) \rightarrow 2V_2O_5 (s) \tag{6.23}$$

Monochrome photographic film consists of an emulsion of silver bromide on a plastic backing. When this is exposed to light, a little of the silver halide decomposes into its elements (Eqn 6.24).

$$2AgBr(s) \rightarrow 2Ag(s) + Br_2(g) \tag{6.24}$$

The 'developer' enhances this process in areas where the film has been exposed to light producing a visible dark 'negative' image of metallic silver. At this stage, however, further exposure would cause the whole film to darken so it is important to remove the unchanged silver halide. The film is 'fixed' by treatment with sodium thiosulphate solution, $Na_2S_2O_3$, which converts the insoluble silver halide into the complex $[Ag(S_2O_3)_2]^{3-}$. Since this has an overall electric charge it is water soluble and is readily washed off the backing, leaving the metallic silver image behind.

Further reading

1. Mingos, D.M.P. (1998). *Essential trends in inorganic chemistry*. Oxford University Press, Oxford, UK.
2. Sanderson, R.T. (1962). *Chemical Periodicity*. Reinhold, New York, USA
3. Mingos, D.M.P. (1995). *Essentials of inorganic chemistry 1*. Oxford University Press, Oxford, UK.
4. Norman, N.J. (1997). *Periodicity and the s- and p-block elements*. Oxford University Press, Oxford, UK.
5. Winter, M.J. (1994). *Chemical bonding*. Oxford University Press, Oxford, UK.
6. Winter, M.J. (1994). *d-Block chemistry*. Oxford University Press, Oxford, UK.
7. WebElements [http://www.webelements.com/] — the periodic table on the WWW.
8. Housecroft, C.E. and Constable, E.C. (1997). *Chemistry – an integrated approach*. Addison Wesley Longman, Essex, UK.
9. Huheey, J.E., Keiter, E.A., and Keiter, R.L. (1993). *Inorganic chemistry – principles of structure and reactivity* (4th edn). Harper International, New York, USA.
10. Shriver, D.F., Atkins, P.W., and Langford, C.H. (1990). *Inorganic chemistry*, Oxford University Press, Oxford, UK.
11. Atkins, P. (1995). *The periodic kingdom*, HarperCollins, New York, USA.
12. Rispoli, P. and Andrew, J. (1999). *Chemistry in focus*. Hodder & Stoughton, UK.
13. Jolly, W.L. (1991). *Modern inorganic chemistry*, McGraw–Hill, Inc., New York, USA
14. Butler, I.S. and Harrod, J.F. (1989). *Inorganic chemistry – principles and applications*, Benjamin/Cummings Publishing Co., Inc., Redwood City, California, USA.
15. Rossoti, H. (1998). *Diverse atoms — profiles of the chemical elements*. Oxford University Press, Oxford, UK.
16. Mackay, K.M. and Mackay, R.A. (1989). *Introduction to modern inorganic chemistry* (4th edn). Blackie, London, UK.
17. Porterfield, W.W. (1984). *Inorganic chemistry — A unified approach*. Addison Wesley, Reading, MA, USA.
18. Cotton, F.A., Wilkinson, G. and Gauss, P.L. (1987). *Basic inorganic chemistry* (2nd edn). John Wiley and Sons, New York, USA.
19. Purcell, K.F., and Kotz, J.C. (1985). *Inorganic chemistry* (Int. Edn). Holt Saunders, Japan.

Index

Periodic table of the elements and element atomic weights (adapted from IUPAC 1991 values)

1 IA IA	2 IIA IIA	3 IIIA IIIB	4 IVA IVB	5 VA VB	6 VIA VIB	7 VIIA VIIB	8 VIIIA VIIIB	9 VIIIA VIIIB	10 VIIIA VIIIB	11 IB IB	12 IIB IIB	13 IIIB IIIA	14 IVB IVA	15 VB VA	16 VIB VIA	17 VIIB VIIA	18 VIIIB VIIIA
1 **H** 1.008																	2 **He** 4.003
3 **Li** 6.941	4 **Be** 9.012											5 **B** 10.811	6 **C** 12.011	7 **N** 14.007	8 **O** 15.999	9 **F** 18.998	10 **Ne** 20.180
11 **Na** 22.990	12 **Mg** 24.305											13 **Al** 26.982	14 **Si** 28.086	15 **P** 30.974	16 **S** 32.066	17 **Cl** 35.453	18 **Ar** 39.948
19 **K** 39.098	20 **Ca** 40.078	21 **Sc** 44.956	22 **Ti** 47.88	23 **V** 50.942	24 **Cr** 51.996	25 **Mn** 54.938	26 **Fe** 55.847	27 **Co** 58.933	28 **Ni** 58.693	29 **Cu** 63.546	30 **Zn** 65.39	31 **Ga** 69.723	32 **Ge** 72.61	33 **As** 74.922	34 **Se** 78.96	35 **Br** 79.904	36 **Kr** 83.80
37 **Rb** 85.468	38 **Sr** 87.62	39 **Y** 88.906	40 **Zr** 91.224	41 **Nb** 92.906	42 **Mo** 95.94	43 **Tc** (97.907)	44 **Ru** 101.07	45 **Rh** 102.906	46 **Pd** 106.42	47 **Ag** 107.868	48 **Cd** 112.411	49 **In** 114.818	50 **Sn** 118.710	51 **Sb** 121.757	52 **Te** 127.60	53 **I** 126.904	54 **Xe** 131.29
55 **Cs** 132.905	56 **Ba** 137.327	57–71	72 **Hf** 178.49	73 **Ta** 180.948	74 **W** 183.84	75 **Re** 186.207	76 **Os** 190.23	77 **Ir** 192.22	78 **Pt** 195.08	79 **Au** 196.967	80 **Hg** 200.59	81 **Tl** 204.383	82 **Pb** 207.2	83 **Bi** 208.980	84 **Po** (208.982)	85 **At** (209.987)	86 **Rn** (222.018)
87 **Fr** (223.020)	88 **Ra** 226.025	89–103	104 **Rf** (261)	105 **Db** (262)	106 **Sg** (266)	107 **Bh** (262)	108 **Hs** (265)	109 **Mt** (266)	110 – (271)	111 – (272)	112 – (277)						

57 **La** 138.906	58 **Ce** 140.115	59 **Pr** 140.908	60 **Nd** 144.24	61 **Pm** (144.917)	62 **Sm** 150.36	63 **Eu** 151.965	64 **Gd** 157.25	65 **Tb** 158.925	66 **Dy** 162.50	67 **Ho** 164.93	68 **Er** 167.26	69 **Tm** 168.934	70 **Yb** 173.04	71 **Lu** 174.967
89 **Ac** 227.028	90 **Th** 232.038	91 **Pa** 231.036	92 **U** 238.029	93 **Np** 237.048	94 **Pu** (244.064)	95 **Am** (243.061)	96 **Cm** (247.070)	97 **Bk** (247.070)	98 **Cf** (251.080)	99 **Es** (252.083)	100 **Fm** (257.095)	101 **Md** (258.10)	102 **No** (259.101)	103 **Lr** (262.11)

Notes: Elements for which the atomic weight is contained within parentheses have no stable nuclides and the weight of the longest-lived known isotope is quoted. The three elements Th, Pa, and U do have characteristic terrestrial abundances and these are the values quoted. In cases where the atomic weight is known to better than three decimal places, the quoted values are rounded to three decimal places. The top, numeric, labelling system (1–18) is the current IUPAC convention. The other two systems are less desirable since they are confusing. There is considerable confusion surrounding the Group labels. The designations A and B are completely arbitrary. The first of these (A left, B right) is based upon older IUPAC recommendations and frequently used in Europe. The last set (main group elements A, transition elements B) was in common use in America. For a discussion of these and other labelling systems see: Fernelius, W.C. and Powell, W.H. (1982). Confusion in the periodic table of the elements, *Journal of Chemical Education*, **59**, 504-508.